MBARI Library

SO-AWI-330

DISCARDED

MLML/MBARI Research Library

Oceanography
in the
Next Decade

Building New
Partnerships

Ocean Studies Board
Commission on Geosciences, Environment, and Resources
National Research Council

NATIONAL ACADEMY PRESS
Washington, D.C. 1992

NATIONAL ACADEMY PRESS • 2101 Constitution Ave., N.W. • Washington, D.C. 20418

NOTICE: The project that is the subject of this report was approved by the Governing Board of the National Research Council, whose members are drawn from the councils of the National Academy of Sciences, the National Academy of Engineering, and the Institute of Medicine. The members of the panel responsible for the report were chosen for their special competencies and with regard for appropriate balance.

This report has been reviewed by a group other than the authors according to procedures approved by a Report Review Committee consisting of members of the National Academy of Sciences, the National Academy of Engineering, and the Institute of Medicine.

This work was sponsored by the National Science Foundation, the Office of Naval Research, the National Oceanic and Atmospheric Administration, the National Aeronautics and Space Administration, the United States Geological Survey, the Department of Energy, and the National Research Council (through Program Initiation Funds).

Oceanography in the next decade : building new partnerships / Ocean
    Studies Board, Commission on Geosciences, Environment, and
    Resources, National Research Council.
        p. cm.
    Includes bibliographical references and index.
    ISBN 0-309-04794-3
    1. Oceanography. I. National Research Council (U.S.).
Commission on Geosciences, Environment, and Resources.
GC11.2.024 1992
551.46—dc20   92-34458
                                                          CIP

Copyright 1992 by the National Academy of Sciences. All rights reserved.

Cover: *West Point, Prout's Neck* by Winslow Homer, 1900. Sterling and Francine Clark Art Institute, Williamstown, Massachusetts.

Printed in the United States of America

OCEAN STUDIES BOARD

Carl Wunsch, Massachusetts Institute of Technology, *Chairman*
Donald F. Boesch, University of Maryland
Peter G. Brewer, Monterey Bay Aquarium Research Institution
Kenneth Brink, Woods Hole Oceanographic Institution
Robert Cannon, Stanford University
Sallie W. Chisholm, Massachusetts Institute of Technology
Biliana Cicin-Sain, University of Delaware
Robert Detrick, Woods Hole Oceanographic Institution
Craig Dorman, Woods Hole Oceanographic Institution
Gordon Eaton, Lamont-Doherty Geological Observatory
Edward A. Frieman, Scripps Institution of Oceanography
Arnold L. Gordon, Columbia University
Gordon Greve, Amoco Production Company
William Merrell, Texas A&M University
Arthur R. M. Nowell, University of Washington
Dennis A. Powers, Stanford University
Brian Rothschild, University of Maryland
John G. Sclater, Scripps Institution of Oceanography
Karl K. Turekian, Yale University

Liaison Members

Robert Beardsley, Commission on Geosciences, Environment,
    and Resources, Woods Hole Oceanographic Institution
Syukuro Manabe, Commission on Geosciences, Environment,
    and Resources, National Oceanic and Atmospheric
    Administration/Geophysical Fluid Dynamics Laboratory
John Orcutt, Chairman, OSB Navy Committee, Scripps
    Institution of Oceanography

Staff

Mary Hope Katsouros, *Director*
Edward R. Urban, Jr., *Staff Officer*
Robin Rice, *Staff Associate*
David Wilmot, *Sea Grant Fellow*
Maureen Hage, *Administrative Assistant*
LaVoncyé Mallory, *Senior Secretary*
Stephen Latham, *Secretary*

The National Academy of Sciences is a private, nonprofit, self-perpetuating society of distinguished scholars engaged in scientific and engineering research, dedicated to the furtherance of science and technology and to their use for the general welfare. Upon the authority of the charter granted to it by the Congress in 1863, the Academy has a mandate that requires it to advise the federal government on scientific and technical matters. Dr. Frank Press is president of the National Academy of Sciences.

The National Academy of Engineering was established in 1964, under the charter of the National Academy of Sciences, as a parallel organization of outstanding engineers. It is autonomous in its administration and in the selection of its members, sharing with the National Academy of Sciences the responsibility for advising the federal government. The National Academy of Engineering also sponsors engineering programs aimed at meeting national needs, encourages education and research, and recognizes the superior achievements of engineers. Dr. Robert M. White is president of the National Academy of Engineering.

The Institute of Medicine was established in 1970 by the National Academy of Sciences to secure the services of eminent members of appropriate professions in the examination of policy matters pertaining to the health of the public. The Institute acts under the responsibility given to the National Academy of Sciences by its congressional charter to be an adviser to the federal government and, upon its own initiative, to identify issues of medical care, research, and education. Dr. Kenneth I. Shine is president of the Institute of Medicine.

The National Research Council was organized by the National Academy of Sciences in 1916 to associate the broad community of science and technology with the Academy's purposes of furthering knowledge and advising the federal government. Functioning in accordance with the general policies determined by the Academy, the Council has become the principal operating agency of both the National Academy of Sciences and the National Academy of Engineering in providing services to the government, the public, and the scientific and engineering communities. The Council is administered jointly by both Academies and the Institute of Medicine. Dr. Frank Press and Dr. Robert M. White are chairman and vice-chairman, respectively, of the National Research Council.

COMMISSION ON GEOSCIENCES, ENVIRONMENT, AND RESOURCES

M. Gordon Wolman, The Johns Hopkins University, *Chairman*
Robert C. Beardsley, Woods Hole Oceanographic Institution
B. Clark Burchfiel, Massachusetts Institute of Technology
Peter S. Eagleson, Massachusetts Institute of Technology
Helen M. Ingram, University of Arizona
Gene E. Likens, New York Botanical Garden
Syukuro Manabe, National Oceanic and Atmospheric
    Administration/Geophysical Fluid Dynamics Laboratory
Jack E. Oliver, Cornell University
Philip A. Palmer, E.I. du Pont de Nemours & Company
Frank L. Parker, Vanderbilt University
Duncan Patten, Arizona State University
Maxine L. Savitz, Allied Signal Aerospace Company
Larry L. Smarr, University of Illinois, Urbana-Champaign
Steven M. Stanley, The Johns Hopkins University
Crispin Tickell, Green College at the Radcliffe Observatory,
    Oxford
Karl K. Turekian, Yale University
Irvin L. White, Battelle Pacific Northwest Laboratories

Staff

Stephen Rattien, *Executive Director*
Stephen D. Parker, *Associate Executive Director*
Janice E. Mehler, *Assistant Executive Director*
Jeanette Spoon, *Administrative Officer*
Carlita Perry, *Administrative Assistant*
Robin Lewis, *Senior Project Assistant*

NATIONAL RESEARCH COUNCIL

2101 CONSTITUTION AVENUE   WASHINGTON, D. C. 20418

OFFICE OF THE CHAIRMAN

The ocean has always had a profound influence on human life and activities. It has been an important source of food and means of commerce. However, it has also been a threat to human life as a focus of war and through its encroachment onto land. In recent decades, the United States has been the world leader in ocean research, both in basic studies and research on the ocean's practical influence on human activities. This pioneering work has largely been the result of remarkably successful partnerships between Federal agencies and research in universities, in which federal agencies support the research of academic scientists and academic scientists provide advice on internal and external research by a variety of mechanisms.

However, the world in which these partnerships were created and sustained is changing rapidly. Concerns about the ocean as a medium for warfare and as a threat to national security are decreasing while environmental problems of the coastal zone and understanding how the ocean controls climate are of increasing importance. Also, major advances in understanding the ocean and in the development of technologies for observing it have set the stage for much greater research achievements. But the potential for such achievement must be set against the human and financial costs of sustaining science. For we are now in a period in which the importance of better understanding the ocean is ever more clear while the resources necessary to obtain this understanding are increasingly scarce.

To understand better what types of partnerships would best serve the United States in the years to come, the Ocean Studies Board of the National Research Council undertook a study of where marine science stands today, how we arrived in this position, and where marine science and technology appear to be headed. This report establishes a framework, in which improved partnerships between the federal government and academic researchers can sustain the advances of the past, and lead our country and the world to greater understanding of the many roles the ocean plays in human life. In their report, the Ocean Studies Board recognizes the ever-growing urgency of the applications but also emphasizes the importance of maintaining the health of the basic science on which all policy decisions must be ultimately based. Although obtaining the proper balance in research funding is essential to national security--in its broad sense--it will not be easy. The Board recommends the use of series of coordinated federal-academic partnerships to achieve a balance in funding among the agencies and a corresponding vitality in basic and applied ocean research.

THE NATIONAL RESEARCH COUNCIL IS THE PRINCIPAL OPERATING AGENCY OF THE NATIONAL ACADEMY OF SCIENCES AND THE NATIONAL ACADEMY OF ENGINEERING

TO SERVE GOVERNMENT AND OTHER ORGANIZATIONS.

This report is the result of the work of many groups and individuals who served as participants in the study, writers of drafts of the report, or as reviewers. The study was begun under the direction of John Sclater, then chairman of the Ocean Studies Board, and completed under the OSB chairmanship of Carl Wunsch. William Merrell chaired this study during its final 9 months, seeing the report to its completion. I thank the chairmen, Board members, and other participants for their efforts in producing a comprehensive report on the future of marine sciences and technology in the United States. I find the call for new and improved partnerships between federal agencies and academia especially timely and believe that the report provides a solid base for building future programs in marine sciences.

In 1991, oceanography reached the end of an era when two major figures passed from the scene. The Ocean Studies Board dedicates this report to Roger Revelle and Henry Stommel, who recognized decades ago the kind of science that would be needed to understand the biology, chemistry, and physics of the ocean and their impact on the Earth system.

Frank Press
Chairman
National Research Council

Preface

The field of oceanography has existed as a major scientific discipline in the United States since World War II, largely funded by the federal government. In this report, the Ocean Studies Board documents the state of the field of oceanography and assesses the health of the partnership between the federal government and the academic oceanography community.

The objectives of the report are to document and discuss important trends in the human, physical, and fiscal resources available to oceanographers, especially academic oceanographers, over the last decade; to present the Ocean Studies Board's best assessment of scientific opportunities in physical oceanography, marine geochemistry, marine geology and geophysics, biological oceanography, and coastal oceanography during the upcoming decade; and, to provide a blueprint for more productive partnerships between academic oceanographers and federal agencies.

The study's approach was to document trends in resources over the past ten to twenty years and to speculate on the likely directions of oceanographic science over the next decade. The board used a number of means to gather information from ocean scientists and from the agencies that conduct and fund ocean science. A meeting on the topic of oceanographic facilities was attended by a number of agency representatives to discuss agency provision and use of these facilities. In addition, agencies that employ oceanographers were surveyed to determined the human resources characteristics of the federal agencies now and over the past twenty years.

A similar questionnaire was sent to the institutional members of the Council on Ocean Affairs to assess the characteristics of the academic oceanography community. Agency and academic scientists were brought together on a number of occasions to discuss the resources available to the field and the science directions of the field.

The board convened two meetings on the future of oceanographic science, one on the West Coast and one on the East Coast. These meetings brought together groups of oceanographers balanced by scientific discipline and these meetings were used to discuss working documents on the exciting science directions of each of the disciplines. The Ocean Studies Board is grateful to those who took on the task of gathering the information for each discipline: Arnold Gordon (physical oceanography), Paul J. Fox and Charles Langmuir (marine geology and geophysics), John Edmond and John Hedges (chemical oceanography and marine chemistry), James Yoder (biological oceanography), and Kenneth Brink (coastal oceanography). Many other people too numerous to cite individually assisted in various aspects of the study, particularly board members who reviewed sections within their fields of expertise.

The board presented its preliminary findings at two meetings of the American Geophysical Union (AGU), one of the major scientific societies to which oceanographers belong. The purpose of the AGU special sessions was to get feedback from scientists in the field of oceanography, to ensure that the board's views were representative of the field as a whole. The first special session, on the topic of resources for oceanography, was held at the December 1990 AGU meeting in San Francisco, California. The second special session, which summarized the results of the two meetings on future oceanographic science, was held at the May 1991 AGU meeting in Baltimore, Maryland.

This study was funded by the agencies that support research in oceanography, including the National Science Foundation, the Office of Naval Research, the National Oceanic and Atmospheric Administration, the Department of Energy, the U.S. Geological Survey, and the National Aeronautics and Space Administration. The board gratefully acknowledges the efforts of the staff of the Ocean Studies Board who labored with the board to produce this report, particularly Mary Hope Katsouros, Edward R. Urban, Jr., Rebecca Metzner Seter, Robin Rice, and David Wilmot. The editor of the report was Sheila Mulvihill.

William J. Merrell
Study Chair

Contents

Oceanography
in the
Next Decade

The Federal Government and academic institutions have together built a research enterprise that is without peer in the world. This enterprise has been based on the concept of a partnership, where each partner contributes and each benefits. But, as in any partnership, a periodic and thorough reexamination is both healthy and necessary, if only to revalidate the original conditions of the partnership.

(p. 138 in Office of Management and Budget, 1992)

Summary

Federal agencies and the academic oceanography community have been fortunate to work together in productive partnerships. These mutually beneficial partnerships are characterized by the federal agencies' funding research at academic institutions that is important to the agencies' missions or is critical to maintaining the health of the basic ocean research endeavor.

These partnerships are likely to change because oceanography is developing a new focus as the results of oceanographic research become increasingly relevant to social and economic concerns. There is an increasing emphasis on global-scale and multidisciplinary research, and a changing mission profile of naval oceanography. Ocean research programs that developed primarily from scientific curiosity have attained increased social meaning and urgency, and federal agencies are increasingly pressured to produce cogent policy options. Yet, over the past decade, academic oceanographers have had access to increasingly limited resources compared to their overall capacity to conduct scientific research. The number of Ph.D.-level academic oceanographers increased dramatically between 1980 and 1990. Also, more sophisticated instrumentation and improved data handling and computing techniques have increased both the scientific capacity of each researcher and the cost of each investigator's research. The net result is a serious imbalance between what can be accomplished and the available resources. Fiscal support in the United States has not kept up with scientific progress, whereas other countries have increased their capacities to conduct oceanographic research. To respond to these challenges, federal agencies and the academic oceanography community need to establish productive new partnerships. Key elements in such partnerships are encouraging individual scientists to take intellectual risks in advancing basic knowledge, providing support that is tied to solving present problems, and encouraging scientists to cooperate in the development of large shared research endeavors. These new partnerships will be the basis of a national oceanographic effort that balances the necessity for a robust program in basic research against the need for research directed at important societal problems.

This report has three major objectives. The first is to document and discuss important trends in the human, physical, and fiscal resources available to oceanographers, especially academic oceanographers, over the last decade. The second goal is to present the Ocean Studies Board's best assessment of the scientific opportunities in physical oceanography, marine geochemistry, marine geology and geophysics, biological oceanography, and coastal oceanography during the upcoming decade. The third and principal objective is to provide a blueprint for more productive partnerships between academic oceanographers and federal agencies. The board attempts to do this by developing a set of general principles that should provide the basis for building improved partnerships and by discussing critical aspects of the specific partnerships for each federal agency with a significant marine program.

OCEANOGRAPHY AND SOCIETY

The ocean dominates Earth's surface and greatly affects daily life. It regulates Earth's climate, plays a critical role in the hydrological cycle, sustains a large portion of Earth's biodiversity, supplies food and mineral resources, constitutes an important medium of national defense, provides an inexpensive means of transportation, is the final destination of many waste products, is a major location of human recreation, and inspires our aesthetic nature.

Today's sense of urgency about ocean studies is precipitated by human impacts on oceanic systems and the need for a better understanding of the ocean's role in controlling global chemical, hydrological, and climate processes. The nation is faced with pressing marine research problems whose timely solution will require increased cooperation between federal agencies and academic scientists. Many of these problems arise from the need to accommodate multiple uses of the ocean and from the ever-increasing concentration of the U.S. population near our coasts. Oceanographic research is important to many of the nation's social concerns, including the following:

• *Global Change*. The ocean is key to regulating both natural and human-induced changes in the planet. The role of ocean circulation and the coupling of the ocean and atmosphere are basic to understanding Earth's changing climate. Regional events such as El Niño and ocean margin and equatorial upwelling influence climate on both seasonal and longer time scales. Earth's population is now large enough to alter the chemical composition of the ocean and atmosphere and to impact the biological composition of Earth.

• *Biodiversity*. The ocean comprises a large portion of Earth's biosphere. It hosts a vast diversity of flora and fauna that are critical to Earth's biogeochemical cycles and that serve as an important source of food and pharmaceuticals. In addition to the exciting discoveries of previously unknown biota near hydrothermal vents, many deep-ocean organisms have evolved under relatively stable conditions. Their unique physiologies and biochemistries have not yet been explored adequately, and methods for sampling the more fragile of these species have been developed only in the past decade. Human influence on marine biota has increased dramatically, threatening the stability of coastal ecosystems. Some species have been overharvested; others have been transported inadvertently to areas where they are not indigenous, sometimes resulting in deleterious effects on native species. Still other species are being cultivated commercially, and aquaculture facilities along coastlines are becoming commonplace in some countries. A better understanding of the ecology of marine organisms is urgently needed to prevent irreversible damage to this living resource.

• *Environmental Quality*. Waste disposed of in coastal areas has reached the open ocean, with broad ramifications for living resources. This problem is compounded because many marine species harvested for commercial and recreational purposes spend a portion of their lives in coastal waters and estuaries. Thus, local pollution can have far-reaching effects.

• *Economic Competitiveness*. Economic prosperity in a global marketplace depends increasingly on technical and scientific applications. There is concern about the ability of the United

States to compete with Europe and Asia. Basic and applied research in the marine sciences and engineering is necessary to achieve and maintain a competitive position in a host of fields, including marine biotechnology, aquaculture, hydrocarbon and mineral exploration and production, maritime transportation, fisheries, treatment and disposal of waste, and freshwater extraction.

• *National Security.* Unprecedented world political changes are redefining national defense interests and altering research and development priorities. Knowledge of the ocean, especially the acoustic properties of marginal seas and coastal areas, is critical to national defense. Experience gained in 1991 during the war in the Persian Gulf highlights the need for better information related to oceanic and coastal processes and to maritime operations and transportation.

• *Energy.* The ocean's energy resources are essential to the national economy and national security. After a decade of relative neglect, energy issues are reemerging. With oil supplies continually threatened by instability in the Middle East and with increasing atmospheric carbon dioxide viewed as a possible trigger of global warming, there is a need to look carefully at a full range of energy sources, from oil and gas in our Exclusive Economic Zone to wave and tidal power and ocean thermal energy conversion. Better knowledge of the ocean and seabed is necessary to exploit responsibly the ocean's untapped petroleum and natural gas resources.

• *Coastal Hazards.* This nation must improve its prediction and response to coastal hazards, both natural and human induced. Hurricanes Hugo and Andrew, two of the strongest hurricanes of the century, devastated parts of the U.S. East Coast. Their impact reinforced the need for better predictive capabilities and a better understanding of coastal storm surges, flooding, erosion, and winds. The exploration for, and production of, petroleum and the transportation of petroleum and chemical products pose risks to the environment when spillage occurs. The movement, effects, and ultimate fates of spilled products must be understood for effective public response. The available information is woefully inadequate, particularly for fragile ecosystems such as coral reefs.

Policy decisions concerning these and many other marine research issues require a comprehensive understanding of the science and engineering of the ocean. Federal, state, and local policies should be based on the best available knowledge of how ocean systems work—their biology, chemistry, geology, and physics. Research results must be communicated effectively to policy makers, with gaps and uncertainties stated clearly and fairly. Also, basic understanding must continue to improve.

MAINTAINING EXCELLENCE

Since World War II, the United States has been the world leader in oceanographic research. Maintaining this excellence requires a talented population of scientists, an informed and educated public, a society interested in and appreciative of new discoveries, open lines of communication between oceanographers and the scientific community at large, and the economic resources necessary to conduct oceanographic research. Continued excellence in oceanography is essential to our national interests and requires constant improvement of both physical and human resources at academic oceanographic institutions. Solving both short- and long-term societal and environmental problems will require well-trained and dedicated scientists working in modern, well-equipped institutions, with sufficient funding. It is critical to the vitality of the ocean enterprise to continue nurturing the academic research environment in which students learn by performing research under the guidance of professors at the forefront of oceanographic science and engineering.

FUTURE OF OCEANOGRAPHIC SCIENCE

Oceanography, the science of the sea, serves many purposes while deriving its impetus from many sources. All of oceanography—physical, chemical, geological, and biological—is driven by scientists interested in advancing basic knowledge. During the past 30 years, marine scientists have verified that Earth's crust is divided into moving plates created at mid-ocean ridges and recycled back into Earth's interior at subduction zones. More recently, dense colonies of animals and bacteria have been discovered at some deep-sea hydrothermal vents and hydrocarbon seeps in ecosystems that only indirectly depend on energy from the Sun. Satellite observations have made possible global estimates of important ocean parameters, such as primary productivity. Our knowledge of interannual climate variations has improved to the point that scientists are now able to forecast El Niño climate disturbances months in advance. These are but a few of the discoveries that have characterized oceanography since the Second World War.

Over the next decade, the field will continue to provide exciting discoveries that contribute to an understanding of Earth as an integrated system and help unravel how humankind may be altering the system. It is now essential and possible to study marine processes on a global scale. Progress in oceanography over the next decade will occur both in the traditional marine science disciplines and, significantly, at the fringes and intersections of these disciplines. Multidisciplinary approaches will lead to new discoveries regarding the ocean's role in climate change, the hydrodynamics of mid-ocean ridges, and the dynamics of coastal processes. Comprehensive study of these topics will require unprecedented levels of cooperation among scientists from numerous disciplines.

Oceanographic studies in the coming decade will focus on how ecosystems affect global cycles of important elements and how changes in the global environment affect marine ecosystems. Studies of the planktonic food web in the sunlit surface waters will advance our understanding of such diverse issues as the role of the ocean in the global carbon cycle and the sustainable yields of commercial fisheries. Studies of ecosystems at deep-sea hydrothermal vents and hydrocarbon seeps will improve theories of the conditions under which life is possible and of the origins of life. More of the ocean will be explored to estimate the extent and nature of deep-ocean vents and their importance in global cycles. Continued study of the ocean's chemistry should bring new understanding of the past state of Earth, and of how marine processes operate today. The study of deep-ocean sediment cores will provide more information about past natural cycles of Earth's climate, against which present climate fluctuations can be calibrated. A better understanding of variability of the circulation of the world ocean, which transports water from near the ocean's surface to deep oceans and back again, will improve our understanding of the variability of the transport of surface water to depth and the interactions with climate.

The foundation of oceanographic knowledge now used in making policy decisions was gained largely through investments in basic research over the past four decades. Oceanographers are privileged to participate in a science that is intellectually compelling and has immediate and long-term practical applications. Yet, the pressure for quick answers to practical questions sometimes obscures the need for investing in the improvement of basic science, which remains the key to solving long term practical problems. Under pressure to provide immediate solutions, it is tempting for agencies whose focus is on their responsibilities for regulation and information provision (mission agencies) to concentrate only on these short-term aspects of their missions. Such mission agencies include the Environmental Protection Agency (EPA), U.S. Geological Survey (USGS), Department of Energy (DOE), National Aeronautics and Space Administration (NASA), parts of the National Oceanic and Atmospheric Administration (NOAA), and others, distinct from the longer-term focus of the National Science Foundation (NSF) and Office of Naval Research (ONR). However, the continued success of these mission agencies ultimately depends on the results of basic research, as well as the results of applied research directed at specific problems.

CONDUCT OF OCEANOGRAPHIC SCIENCE

In the past decade, oceanography has rapidly incorporated new technologies from other fields, remote sensing, material science, electronics, and computer science, for example. A fundamental change arising from the use of these new technologies is an increase in the quality and quantity of data collected and a dramatic increase in each oceanographer's capacity to study oceanic phenomena. As in many fields, the cost of making new discoveries in oceanography has escalated because these discoveries have been achievable only with the development and use of new satellites, vessels, laboratory and field instruments, and computers. This increased cost translates into an increased cost per scientist in the field, in what has been referred to as the "sophistication factor" by the President's Science Advisor, Dr. D. Allan Bromley. Yet, when adjusted for inflation, total research funding for ocean science has remained nearly constant over the past decade. During this interval, the number of Ph.D.-level academic oceanographers has increased by half again. The increase in the scientific capacity of each investigator and in the total number of qualified investigators, coupled with nearly constant overall federal funding, has resulted in inadequate support for many capable researchers.

Another significant change in the past decade is the onset of large-scale, long-term global research programs. Primarily planned and begun with NSF support, these programs focus the work of many scientists on global questions. These large programs are usually managed through national or international consortia that involve many scientists, multiple agencies, and often a number of countries. Such programs will explore new questions and test new mechanisms for working together in the next decade. Uncertainties about the changing environment of the planet are rapidly moving much of oceanography from a focus on projects that use the capabilities and interests of a single investigator for a limited time to projects that require the involvement of many individuals, institutions, and governments for decades. Special attention should be given to integrating mission agencies into the planning and execution of these long-term programs. Mechanisms must be developed to coordinate the plans of foreign nations, federal agencies, academic institutions, and individual scientists, and to sustain these large-scale efforts in a scientifically and technically sound manner.

The realization that global-scale studies are now not only possible but necessary is a major impetus for new partnerships in oceanography. Indeed, the design and deployment of a long-term global ocean observing system, now being planned, will be possible only if such partnerships are realized and the cooperation of marine scientists and governments throughout the world is achieved.

TOWARD NEW PARTNERSHIPS

Traditional partnerships in the ocean sciences have consisted primarily of academic scientists submitting proposals to the National Science Foundation and the Office of Naval Research for funding. This funding system is powerful and flexible, allowing the NSF and ONR to fund excellent scientists whose areas of expertise are those necessary to solve problems at the forefront of oceanography. Through their support of research and related infrastructure, these two agencies sustain the basic research programs at academic oceanographic laboratories. If significant progress in our basic understanding of the ocean is to continue, the excellent relationships of NSF and ONR with the academic community must be maintained. Agencies that fund oceanography can help maintain competence in the field as problem areas change. Flexibility and variety in scientific approaches can be maintained by an extramural funding strategy that both responds to changing problems and needs, and maintains a strong overall base of scientific activities in the field as a whole. It is more difficult for agencies to respond quickly to change through their own laboratories.

Many other federal agencies are also involved in marine science and policy, but their use of the marine science knowledge and their responsibility to the academic community vary widely. Agency responsibilities range from NSF's and ONR's active promotion of the health of basic science to the highly specific and practical rule-making procedures of the Environmental Protection Agency. The National Oceanic and Atmospheric Administration has a wide range of responsibilities in the ocean but is only now beginning to develop significant research programs in many of its areas of responsibility. The future vitality of basic oceanographic research in academia may depend on its forging productive partnerships with NOAA. Partnerships between academic oceanographers and NASA, DOE, USGS, or the Minerals Management Service will add diversity and vitality to the national oceanographic effort.

No simple description can usefully encompass the range of partnerships between federal agencies and the academic oceanography community. However, under the traditional arrangement, mission agencies, such as EPA, have received relatively little intellectual input from academia and provided relatively little funding to academic institutions. These agencies, whose short-term missions often require highly applied research, rely primarily on their own scientists. Yet, these same agencies have relied on academic scientists to provide the underpinning knowledge upon which their policy decisions are based. In general, the mission agencies have not contributed much to advancing fundamental knowledge in their areas of concern, perhaps assuming that NSF or ONR would fund basic research adequately. Such a perspective has the danger of focusing oceanography primarily on short-term applied problems. Achieving a sensible balance between basic and applied oceanographic research should be the concern of each agency using the results of ocean research.

As the context in which oceanography is conducted changes, how can federal agencies and oceanographers in academic institutions strengthen and improve their cooperative efforts? In general, partnerships must be extended beyond financial relationships to include the sharing of intellect, experience, data, instrument development, facilities, and labor.

Communication

Many mission agencies and academic scientists have little experience in interacting with one another, but both groups would benefit from doing so. **The board recommends that each agency with an ocean mission and without existing strong links to the nongovernment community establish permanent mechanisms for ensuring outside scientific advice, review, and interaction.** The obvious advantage of external consultation is that it provides an objective evaluation of agency needs and poses possible solutions from a new perspective. The National Research Council is but one possible source of external advice. **These advisory groups should report to a level sufficiently high that their views are presented directly to agency policy makers and the relationships are eventually institutionalized to establish a collective memory.**

The board recognizes that the existence of multiple marine agencies with differing mandates brings a vigor and diversity to the field. However, the lack of coordination and cooperation among agencies that conduct or sponsor marine research detracts from this advantage. Informal attempts at coordination have been largely unsuccessful; a formal mechanism is necessary. **The board recommends that, because no single agency is charged with and able to oversee the total national marine science agenda, an effective means be found for agencies to interact at the policy level and formulate action plans.**

One model for such interaction is the Committee on Earth and Environmental Sciences of the Federal Coordinating Council for Science, Engineering, and Technology. Regardless of the coordinating mechanism chosen, it must permit the agencies to develop a synergistic approach to addressing national problems and to coordinating programs and infrastructure. High-priority tasks for such a group would be an examination of the balance between individual investigator awards and large project support, and the establishment of guidelines for the large, global change projects.

Agency Responsibility for Basic Science

The vitality of basic ocean research in the United States resides principally in its academic institutions. **The board recommends that federal agencies with marine-related missions find mechanisms to guarantee the continuing vitality of the underlying basic science on which they depend.** In some agencies, the best mechanism is direct funding of individual investigator grants; in others, consultation and collaboration work well. NSF and, secondarily, ONR should retain primary responsibility for the vitality of the basic science, with NOAA becoming increasingly involved. Also, mission agencies such as EPA and DOE must share more fully in this responsibility. It is particularly important to encourage involvement of mission agencies in sampling and monitoring programs pertaining to long-term global change issues. At present, a disproportionate share of the funds is provided by NSF. As these programs expand, resources for individual investigator grants could be reduced if other agencies do not assume responsibility for some of the funding.

Responsibility of Academic Institutions

Through the years, academic oceanographic institutions evolved different organizational structures ranging from typical academic departments to large comprehensive institutions that operate multiple ships and shared facilities. As the benefits of cooperation became evident, vehicles for the cooperative use of ships and some other facilities have developed. **The board recommends that academic oceanographic institutions find additional ways to achieve cohesiveness among these institutions and a sense of common scientific direction.** It is essential that this cooperation be achieved at both the administrative and the working-scientist levels so that the interactions are based on the needs of science as well as the needs of the institutions. **The board also recommends that the academic institutions, individually or through consortia, take a greater responsibility for the health of the field, including nationally important programs.** In particular, the large, long-lived global change research programs are indicative of the need for institutional responses that are of longer duration and more stable than those of individual scientists. Also, the heavy dependence of academic oceanographers on federal support, compared with other fields, suggests that the academic institutions should explore mechanisms for the stable support of academic researchers. Academic scientists have a responsibility to help the federal agencies that fund them when it comes to applying research results to agency missions. Partnerships imply shared responsibilities and anticipation of the future needs of both partners.

Sharing of Academic and Federal Resources

The board recommends that federal and academic researchers improve the sharing of data, the cooperative use of facilities, and the conduct of joint research. Some mission agencies encourage cooperation with academic scientists, but increased formal interaction could significantly improve the efficiency of the national oceanographic effort. The major facility available to the marine science community, the research fleet, is a national resource. Maintaining, developing, and operating the fleet in the most efficient and cost-effective manner should be paramount in all discussions of shared resources.

Development of Instrumentation

Some advancement of oceanographic knowledge has come through the development of new observational technologies. Effective operational systems to solve the complex problems facing mission agencies will consist largely of instruments that either do not now exist or have not yet been redesigned for oceanography. The development of both in situ and satellite oceanographic instrumentation requires a long-term investment in novel technologies and in the extensive field trials necessary to make instruments operational. **The board recommends that to ensure continued progress in instrumentation, new mechanisms be found to address the long time frames necessary for instrument development in oceanography.** Mission agencies, whose success will depend increasingly on instrumentation that does not yet exist, should initiate suitable roles in the development of new technology.

Transfer of Responsibility

The division of tasks between academic scientists and agencies will depend on the agencies' missions, resources, and internal capabilities vis-à-vis the academic community's. Mechanisms must be developed to provide smooth transition from research activities to operational measurements. In particular, the proposed global ocean observing system will necessitate unprecedented levels of monitoring. **The board recommends that academia and federal agencies work together to ensure that appropriate long-term measurements are extended beyond the work of any individual scientist or group of scientists and that the quality of such measurements is maintained.**

Data Management and Exchange

The board recommends that the present system for data management and exchange within and among the various elements of the marine science community be modernized to reflect the existence of distributed computing systems, national and international data networks, improved satellite data links, and on-line distribution of oceanographic data. Also, provision must be made for future access to existing data.

Specific Partnerships

These general recommendations form the basis for building new partnerships between federal and academic interests in ocean science. Of course, they do not apply to all agencies to the same degree. The full report discusses aspects of specific partnerships for federal agencies with significant ocean programs. The board believes that if these new partnerships are established and nurtured, the next decade of ocean science research will be characterized by a robust program of basic research and significant progress toward the solution of marine problems of importance to humankind.

1

Introduction

IMPORTANCE OF THE OCEAN TO SOCIETY

The ocean dominates Earth's surface and greatly affects our daily lives. It regulates Earth's climate, plays a critical role in the hydrological cycle, sustains a large portion of Earth's biodiversity, supplies food and mineral resources, constitutes an important medium of national defense, provides an inexpensive means of transportation, is the final destination of many waste products, is a major location of human recreation, and inspires our aesthetic nature.

Today's sense of urgency about ocean studies is precipitated by human impacts on oceanic systems and the need for a better understanding of the ocean's role in controlling global chemical, hydrological, and climate processes. The nation is faced with pressing marine research problems whose timely solution will require increased cooperation between federal agencies and academic scientists. Many of these problems arise from the need to accommodate multiple uses of the ocean and from the ever-increasing concentration of the U.S. population near its coasts. Oceanographic research is important to many of the nation's social concerns, including the following:

• *Global Change.* The ocean plays a predominant role in regulating both natural and human-induced changes in our planet. The role of ocean circulation and the coupling of the ocean and the atmosphere are basic to understanding Earth's changing climate. Regional events such as El Niño and ocean margin and equatorial upwelling influence climate on both seasonal and longer time scales. The world's population is now large enough to alter the chemical composition of the ocean and atmosphere and to impact the biological composition of Earth.

• *Biodiversity*. The oceans comprise a large portion of Earth's biosphere and support a vast diversity of flora and fauna that are critical to Earth's biogeochemical cycles and serve as an important source of food and pharmaceuticals. In addition to the exciting discoveries of previously unknown biota near hydrothermal vents, many deep-ocean organisms have evolved under relatively stable conditions. Their unique physiologies and biochemistries have not yet been explored adequately, and methods for sampling the more fragile of these species have been developed only in the past decade. Human influence on marine biota has increased dramatically, threatening the stability of coastal ecosystems. Some species have been overharvested; others have been transported inadvertently to areas where they are not indigenous, sometimes resulting in deleterious effects on native species. Still other species are being cultivated commercially, and aquaculture facilities along coastlines are becoming commonplace in some countries. A better understanding of the ecology of marine organisms is urgently needed to prevent irreversible damage to this living resource.

• *Environmental Quality*. Waste disposed in coastal areas has reached the open ocean, with broad ramifications for living resources. This problem is compounded because many marine species harvested for commercial and recreational purposes spend a portion of their lives in coastal waters and estuaries. Thus, local pollution can have far-reaching effects.

• *Economic Competitiveness*. Economic prosperity in a global marketplace depends increasingly on technical and scientific applications. There is concern about the ability of the United States to compete with Europe and Asia. Basic and applied research in marine science and engineering is necessary to achieve and maintain a competitive position in a host of fields, including marine biotechnology, aquaculture, hydrocarbon and mineral exploration and production, maritime transportation, fisheries, treatment and disposal of waste, and freshwater extraction.

• *National Security*. Unprecedented world political changes are redefining national defense interests and altering research and development priorities. Knowledge of the ocean, especially the acoustic properties of marginal seas and coastal areas, is critical to national defense. Experience gained in 1991 during the war in the Persian Gulf highlights the need for better information related to oceanic and coastal processes and to maritime operations and transportation.

• *Energy.* The ocean's energy resources are essential to the national economy and national security. After a decade of relative neglect, energy issues are reemerging. With oil supplies continually threatened by instability in the Middle East and with increasing atmospheric carbon dioxide viewed as a possible trigger of global warming, there is a need to look carefully at a full range of energy sources, from oil and gas in our Exclusive Economic Zone to wave and tidal power and ocean thermal energy conversion. Better knowledge of the ocean and seabed is necessary to exploit responsibly the ocean's untapped petroleum and natural gas resources.

• *Coastal Hazards.* This nation must improve its prediction of and response to coastal hazards, both natural and human induced. Hurricanes Hugo and Andrew, two of the strongest hurricanes of the century, devastated parts of the U.S. East Coast. Their impact reinforced the need for better predictive capabilities and a better understanding of coastal storm surges, flooding, erosion, and winds. The exploration for, and production of, petroleum and the transportation of petroleum and chemical products pose risks to the environment when spillage occurs. The movement, effects, and ultimate fates of spilled products must be understood for effective public response. The available information is woefully inadequate, particularly for fragile ecosystems such as coral reefs.

Increasing our knowledge about the ocean is a matter of urgency. Human-induced changes to the planet's oceans and atmosphere will increasingly affect the global cycles that ultimately control the number of people our planet can support. To predict the results of environmental disturbances and prescribe possible remedies, a better understanding of Earth's systems, including the ocean, must be acquired. For example, an important scientific and policy question today is whether Earth will warm in response to increasing concentrations of greenhouse gases in the atmosphere and, if so, how quickly. We know that the concentrations of these gases are increasing and that the most advanced climate models indicate that warming should occur. The ocean plays a key but poorly understood role in moderating both greenhouse gases and temperature change.

The coast of the United States is one of the nation's most

valuable geographic features. It is at the junction of land and sea that most of the nation's trade and industry take place. The effectiveness with which the resources of the coastal zone are used is a matter of national importance. The multiple uses of valuable coastal areas generate intense state and local interest. From 1950 to 1984 the population in coastal counties grew more than 80 percent. By 1995, more than three-fourths of the U.S. population will live within 50 miles of the coastline.

Coastal waters and estuaries provide food and are the shelter and spawning grounds for almost two-thirds of the nation's commercial fish stocks. Oil, gas, and mineral resources in the coastal waters are essential to our national economy and security. Since the first offshore oil well was drilled off California in 1896, numerous oil and gas pools have been discovered near our coasts.

Recent reports of increased pollution of estuarine and coastal waters are cause for serious concern and action. Waste disposal, especially from pipelines, runoff, and dumping at sea, jeopardizes our ocean and coastal waters. The toll that waste takes on the ocean is persistent and growing. The continuing damage to estuarine and nearshore resources from pollution, development, and natural forces raises serious doubts about the survival of these systems. Better understanding of these systems is essential for good policy decisions.

Policy decisions concerning these and many other interactions of the ocean with everyday life rest upon a sound scientific understanding of the ocean. To the extent that such policy decisions are to be useful, they must be consistent with the best available information about how the system works: its physics, chemistry, geology, and biology. Both the government and the scientific community as a whole must ensure that what is known about the ocean is made available to policy makers, that what is not known is clearly stated, and that progress in furthering our basic understanding continues.

MAINTAINING EXCELLENCE

Our nation excels in oceanography. Since World War II, the United States has been a world leader in essentially every area of oceanography. To maintain this excellence will require a talented population of scientists, an informed and educated public, a society that is interested in and appreciative of new discoveries, open lines of communication between oceanographers and the scientific community at large, and economic resources for conducting oceanographic research on the frontiers of knowledge. Excellence in oceanography also requires harmony between its basic scientific aims and the pressing needs of society.

We cannot take for granted the continued excellence of oceanography in the United States because the foundation of facilities and human resources developed in the past must be renewed constantly. Continued excellence in oceanography is essential to the national interests of the United States. Agencies that fund oceanography can help maintain the competence of the field as problem areas change. Flexibility and variety in scientific approaches can be maintained by an extramural funding strategy that both responds to changing problems and needs and maintains a strong overall base of scientific activities in the field as a whole. It is more difficult for agencies to respond quickly to change through their own laboratories.

Vannevar Bush's *Science: The Endless Frontier* is still the classic statement of the essential ingredients of scientific excellence. He noted that "without scientific progress no amount of achievement in other directions can insure our health, prosperity, and security as a nation in the modern world. This essential new knowledge can only be obtained through basic scientific research." He further stated that "basic research is performed without thought of practical ends . . . leads to new knowledge, provides scientific capital, creates the funds from which the practical applications of knowledge must be drawn." Finally, he stated that "government must fund science in accordance to certain fundamental principles" including the essentiality of quality, improved efficiency of research expenditures, and increased cooperation in setting goals and priorities.

U.S. OCEANOGRAPHY SINCE WORLD WAR II

In the aftermath of World War II, the United States constructed a scientific research mechanism of outstanding success, which for years dominated scientific progress. Many studies described the nature of this research enterprise. A wide consensus exists that much of its success has been due to the partnership between the federal agencies that became the patrons of science and technology and the major research universities, both public and private. Marine science shared in the general outstanding progress, although its history is exceptional in several ways.

The war thrust the United States into global affairs, and its many sea campaigns not only drew public interest to the ocean but also highlighted our ignorance of it. Most members of the small marine science community turned to military-oriented work in uniform, in the civil service, or at universities and related institutions. Academic ships, as well as those of the federal government, were put to work on Navy research and surveying tasks. The Navy needed and received oceanographic help in everything from submarine warfare to amphibious landings. Although this assistance contributed to the war effort, of even more importance, it impressed on the nation the fact that marine science was not an abstract endeavor but could contribute to the public good in many fields.

The plan of Vannevar Bush at the end of World War II for government support of university science led to the formation of the Office of Naval Research (ONR). It was charged with ensuring the development of strong academic research programs in scientific fields of interest to the Navy. The growing Cold War and the threat from both surface and, particularly, submarine vessels led ONR to conclude that expanding and generally strengthening the basic science of the ocean were in the national interest. With ONR's financial backing, existing marine research centers were expanded and new ones created. Initially, ONR was more concerned with institutional support than with program definition. There was generally only one contract per institution, proposals of work were often loosely defined, and the director of the institution had considerable discretion in transferring funds from one investigator to another. In 1950, the National Science Foundation (NSF), dedicated primarily to the support of peer-reviewed single-investigator research in the academic community, was created.

The postwar and post-*Sputnik* periods from 1960 to 1980 were marked by a national awareness of the rest of the world and an intense interest in science. These encouraged international cooperation in research, tempered strongly by a U.S. desire to achieve world leadership in science and technology. In marine science, interest grew from our coastlines to the globe, leading to such major ocean-related programs as the International Geophysical Year, the Deep Sea Drilling Project, and the International Decade of Ocean Exploration. Through both its small science programs and large coordinated programs, NSF rapidly became a significant supporter of oceanography and is now the dominant supporter of academic ocean research. The Navy, which almost single handedly provided impetus and financial support for the postwar academic expansion in oceanography, has progressively concentrated its support in a relatively limited number of Navy-relevant areas and in providing major oceanographic research vessels. NSF has increasingly borne the costs of both research and ship operations.

The National Oceanic and Atmospheric Administration (NOAA), established in 1970, has developed several mechanisms for working with the academic community. NOAA's National Sea Grant College Program added a new dimension to university marine science programs by concentrating primarily on applied coastal research and developing extension and public information networks. In particular, Sea Grant supported areas of marine science not emphasized by ONR and NSF—the study of estuaries, fisheries, and pollution and the transition of such research to practical applications. The proximity of NOAA oceanographic and fisheries laboratories to academic institutions leads to opportunities for joint educational and research programs, of benefit to both the academic and the federal laboratories. NOAA provides comparatively modest extramural research funds as part of its Climate and Global Change Program and its Coastal Ocean Program and through the National Marine Fisheries Service. Other federal agencies support academic scientists, notably the Departments of Interior and Energy, the Environmental Protection Agency, and the National Aeronautics and Space Administration.

In the past decade, oceanography has incorporated new technologies from other fields, for example, space research, electronics, and computer science. A fundamental change arising from the use of new technologies has been an increase in both the quality and the quantity of data collected. Thus each oceanographer's capacity to study ocean phenomena has increased dramatically. This increase has also increased the cost of each oceanographer's scientific research.

Another significant change is the planning, primarily with NSF support, of large-scale, long-term global research programs that focus the work of many scientists on global ocean questions. These large programs are part of the overall scientific quest. They are usually managed by international consortia that involve many scientists, multiple agencies, and often a number of countries. The experience of working in these programs will lead us to ask different questions and to explore different mechanisms of working together in the next decade.

U.S. STYLE OF LARGE PROGRAM MANAGEMENT

Since the 1970s, U.S. marine scientists and the federal government have shown remarkable ingenuity in developing mechanisms to meet the challenges of large new programs. Instead of developing large permanent organizations with new facilities as in some other countries, U.S. programs, such as the Mid-Ocean Dynamics Experiment, Geochemical Ocean Sections, Coastal Upwelling Ecosystem Analysis, and the Climate: Long-Range Investigation, Mapping, and Prediction projects, have evolved differently. Large programs typically developed within the academic community through workshops. The community formed scientific steering groups, which were accepted and funded by NSF and other federal agencies, and set up program offices. These offices are located at academic institutions, and program staff is hired for the project duration. The program office may move as the leadership of the program changes. Upon completion of the research program, the staff assumes other duties and the facilities are used for other purposes, so there is no long-term drain on agency resources.

OCEANOGRAPHIC RESEARCH, NATIONAL AND INTERNATIONAL

Oceanographic research involves studies of the motion of the water, the distribution of marine life, and the interaction of seawater with ocean boundaries. Knowledge of the exchanges of energy, heat, and mass at the ocean-atmosphere interface is important to climate and weather prediction. Oceanographic research has advanced from the past era of exploration to one of observation and description of ocean systems and of processes within the ocean and among the ocean, atmosphere, and ocean basins and boundaries. Because of the advances in satellite observation, computer modeling, and technology (e.g., global positioning systems and acoustic tomography), the coming decade of research holds much promise.

The ocean science community has developed several multi-institutional, interdisciplinary research programs that should significantly improve our knowledge of physical, chemical, geological, and biological processes occurring in the ocean. One important goal of these programs is to understand ocean processes in sufficient detail to allow predictions to be made of the impact of human activities on the environment. Because of the global scale of many environmental problems and the substantial resources (i.e., financial, infrastructure, and human) required, large ocean research programs are often cooperative international efforts.

The nation's academic capability in ocean science is robust. It is reflected in strong academic departments at many public and private universities, mixed with a few large oceanographic centers. U.S. academic oceanographers are internationally recognized leaders who are key to international scientific activity. Although the United States funds perhaps half of the global total of oceanographic research in many of its disciplines, international cooperation is vital for achievement of the goals of most large global research programs. The academic community could contribute significantly to the study of the ocean and to solutions to the spectrum of ocean-related environmental problems now facing the nation and the world.

THIS REPORT

Objectives

This report has three major objectives. The first is to document and discuss important trends in the human, physical, and fiscal resources available to oceanographers, especially academic oceanographers, over the last decade. Its second goal is to present the board's best assessment of the scientific opportunities in physical oceanography, marine geochemistry, marine geology and geophysics, biological oceanography, and coastal oceanography during the upcoming decade. The third and principal objective is to provide a blueprint for more productive partnerships between academic oceanographers and federal agencies. The board attempts to do this by developing a set of general principles that should provide the basis for building improved partnerships and by discussing critical aspects of the specific partnerships for each federal agency with a significant marine program.

Contents

Chapter 1 introduces the importance of the ocean to society and the need for maintaining excellence in marine-related research and education. The growth of U.S. academic oceanography since World War II and the structure of both national and international research are discussed.

Chapter 2 discusses partnerships in ocean science. A general partnership theme is presented, followed by specific partnership possibilities with agencies of the federal government. This report does not discuss partnerships with states and industry, which may be explored by the board at a later time.

Chapter 3 details some of the scientific opportunities of the next decade and some of the most important ongoing research programs. It describes opportunities and programs for each of the four major subdisciplines of oceanography: physical, chemical, geological, and biological, as well as for the interdisciplinary area of coastal oceanography.

Chapter 4 presents information about the infrastructure of oceanography. Included is a discussion of the human, physical, and fiscal resources. This initial overview of the field's resources raises many questions that should be examined at a later date. In-depth analysis and synthesis remain to be carried out.

2

Toward New Partnerships
in Ocean Sciences

Since about 1950, scientific research in the United States has been characterized by federal funding of academic scientists to conduct research of general interest to the government. This defines a partnership of sorts, a mutually beneficial relationship between the federal government and academic scientists. In ocean science to date, these traditional partnerships have consisted primarily of scientists in academic and private institutions submitting proposals to the National Science Foundation (NSF) and the Office of Naval Research (ONR). This funding system is powerful and flexible, allowing NSF and ONR to fund excellent scientists whose areas of expertise are those necessary to solve problems at the forefront of oceanography. The two agencies encourage and sustain basic research programs at academic and private laboratories. The numerous federal agencies involved in marine science and policy differ greatly in their use of marine science knowledge and in their responsibility to the academic community. Agency responsibilities range from NSF's and ONR's active promotion of the health of basic science to highly specific and practical rule-making procedures of the Environmental Protection Agency (EPA). The National Oceanic and Atmospheric Administration (NOAA) has a wide range of responsibilities in ocean matters but is just beginning to develop significant research programs in many of its areas of responsibility. The future vitality of basic oceanographic research within academia may depend on its forging productive partnerships with NOAA. No simple description can usefully encompass the range of partnerships between federal agencies and the academic oceanography community. However, under the traditional arrangement, mission agencies (e.g., EPA) received relatively little direct intellectual input from academic and private scientists, and provided relatively little funding to academic institutions. Yet, although such agencies have relied on academic scientists for much of the basic knowledge required to understand policy questions, they have not assumed a serious responsibility to advance that knowledge. These agencies, whose short-term missions often require applied research, rely primarily on agency scientists to carry out their missions with optimal short-term efficiency.

The traditional scientific partnerships that have existed over the past 40 years are likely to change because the focus of oceanography and the way it is carried out are changing. Increased emphasis on the global scale and on multidisciplinary research, the changing emphasis of naval oceanography, and increasingly limited resources relative to an expanded capacity to conduct science by using modern instrumentation and computing are all contributing to change. These factors are pushing the field of oceanography toward serious consideration of the greater efficiency that could be achieved by a better coordinated national oceanography effort.

Our nation is faced with many pressing problems whose solutions would benefit from increased cooperation between federal agencies and nongovernmental scientists. Ocean research programs that developed from scientists' curiosity about nature have a new social context and urgency. A salient example is global change in all its aspects, including ocean circulation, air-sea transfer of gases, response of organisms, sea-level rise, and other effects of a potentially warming Earth. A balance should be maintained between the complementary approaches of large programs and individual investigator science in order to preserve the diversity and vigor of the field. Individual investigator science can be a fertile source of innovative ideas, whereas large programs can garner the resources for global-scale studies and can add momentum, collective wisdom, and resources for long-range planning.

A major impetus for new partnerships in oceanography is the realization that a global scale of study is now both possible and desirable. The design and deployment of a global ocean observing system, now being discussed, will be possible only with cooperation among the world's ocean scientists and its governments. Such a system will be necessary for obtaining enough long time-series global data to understand the global climate system and predict its response to human influence.

Oceanography is changing rapidly from its focus on the capabilities and interests of single or small groups of investigators involved in studies of limited duration to a focus on scientific questions of global scope, involving large numbers of individuals, institutions, and governments; spanning decades; and having major significance to society. The role of the individual investigator in this context has not lessened. Mechanisms must be developed by which these new large-scale efforts are sustained in a scientifically and technically sound manner and the plans of a variety of federal agencies and nations are coordinated.

A major reason for the preeminence of U.S. marine science is the great diversity of institutions in the field. This diversity is a key to future strength and it needs to be maintained. This statement does not suggest, however, that the present numbers and types of institutions are necessarily optimal for the future.

GENERAL PARTNERSHIP THEMES

The health of the marine sciences in the United States must be maintained because of the continuous need for fundamental knowledge as the basis for developing sound public policy. The health of ocean science depends on a complex symbiosis that must be constantly nurtured. The academic and private oceanographic institutions, working with the federal government, have shown remarkable ingenuity in developing mechanisms to coordinate multi-institutional resources (e.g., the University-National Oceanographic Laboratory System (UNOLS) and the Joint Oceanographic Institutions, Inc. (JOI)). UNOLS is a multi-institution system for coordinating scheduling, safety, refitting, and replacement of academic oceanographic vessels. JOI, governed by representatives of 10 of the largest oceanographic institutions, was founded initially to manage the Deep Sea Drilling Project; JOI now undertakes broader responsibilities for large programs and new technology. In addition, several research programs (e.g., the Tropical Ocean-Global Atmosphere program, Mid-Ocean Dynamics Experiment, Geochemical Ocean Sections, and Coastal Upwelling Ecosystem Analysis) successfully combined the efforts of U.S. government agencies, agencies of other countries, and federal and nongovernmental scientists, both domestically and internationally.

As the context in which oceanography is conducted changes, how can partnerships between federal agencies and oceanographers in academic and private institutions be strengthened and improved? In general, the partnerships must extend beyond financial relationships to include the sharing of intellect, data, instrument development, facilities, and labor. Key elements in such partnerships are encouraging individual scientists to take intellectual risks in advancing basic knowledge, providing support that is tied to solving existing problems, and encouraging scientists to cooperate in the development of large shared research endeavors.

Communication

Many mission agencies and academic scientists have little experience interacting with one another, but both groups would benefit from doing so. **The board recommends that each agency with an ocean mission and without existing strong links to the nongovernment community establish permanent mechanisms for ensuring outside scientific advice, review, and interaction.** The obvious advantage of external consultation is that it provides an objective evaluation of agency needs and poses possible solutions from a new perspective. The National Research Council (NRC) is but one possible source of external advice. **These advisory groups should report to a level sufficiently high that their views are presented directly to agency policy makers and the relationships are eventually institutionalized to establish a collective memory.**

The board recognizes that the existence of multiple marine agencies with differing mandates brings a vigor and diversity to the field. However, the lack of coordination and cooperation among agencies that conduct or sponsor marine research detracts from this advantage. Informal attempts at coordination have been largely unsuccessful; a formal mechanism is necessary. **The board recommends that, because no single agency is charged with and able to oversee the total national marine science agenda, an effective means be found for the agencies to interact at the policy level and formulate action plans.**

One model for such interaction is the Committee on Earth and Environmental Sciences of the Federal Coordinating Council for Science, Engineering, and Technology. Regardless of the coordinating mechanism chosen, it must permit the agencies to develop a synergistic approach to addressing national problems and to coordinating programs and infrastructure. High-priority tasks for such a group would be examination of the appropriate balance between individual investigator awards and large project support and the establishment of guidelines for the large, global change projects.

Agency Responsibility to Basic Science

The vitality of basic ocean research in the United States resides principally in its academic institutions. **The board recommends that federal agencies with marine-related missions find mechanisms to guarantee the continuing vitality of the underlying basic science on which they depend.** In some agencies, the best mechanism is direct funding of individual investigator grants; in others, consultation and collaboration work well. NSF and, secondarily, ONR should retain primary responsibility for the vitality of the basic science, with NOAA becoming increasingly involved. Also, mission agencies such as EPA and the Department of Energy (DOE) must share more fully in this responsibility. It is particularly important to encourage the involvement of mission agencies in sampling and monitoring programs pertaining to long-term global change issues. At present, a disproportionate share of the funds is provided by NSF. As these programs expand, resources for individual investigator grants could be reduced if other agencies do not assume responsibility for some of the funding.

Responsibility of Academic Institutions

Through the years, academic oceanographic institutions evolved different organizational structures ranging from typical academic departments to large comprehensive institutions that operate multiple ships and shared facilities. As the benefits of cooperation became evident, arrangements for the cooperative use of ships and some other facilities have developed. **The board recommends that academic oceanographic institutions find additional ways to achieve cohesiveness among the institutions and a sense of common scientific direction.** It is essential that this cooperation be achieved at both the administrative and the working-scientist levels so that the interactions are based on the needs of science as well as the needs of the institutions. **The board also recommends that academic institutions, individually or through consortia, take a greater responsibility for the health of the field, including nationally important programs.** In particular, the large, long-lived global change programs could benefit from institutional responses that are of longer duration and more stable than those of individual scientists. Also, the heavy dependence of academic oceanographers on federal support, compared with other fields, suggests that academic institutions should explore mechanisms for the stable support of their researchers. Academic scientists have a responsibility to help the federal agencies that fund them when it comes to applying research results to agency missions. Partnerships imply shared responsibilities and anticipation of the future needs of both partners.

Sharing of Academic and Federal Resources

The board recommends that federal and academic researchers improve the sharing of data, the cooperative use of facilities, and the conduct of joint research. Some mission agencies encourage cooperation with academic scientists, but increased formal interaction could significantly improve the efficiency of the national oceanographic effort. The major facility available to the marine science community, the research fleet, is a national resource. Maintaining, developing, and operating the fleet in the most efficient and cost-effective manner should be paramount in all discussions of shared resources.

Development of Instrumentation

Some advancement of oceanographic knowledge has come through the development of new observational technologies. Effective operational systems to solve the complex problems facing mission agencies will consist largely of instruments that either do not now exist or have not yet been redesigned for oceanography. The development of both in situ and satellite oceanographic instrumentation requires a long-term investment in novel technologies and in the extensive field trials necessary to make instruments operational. **The board recommends that to ensure continued progress in instrumentation, new mechanisms be found to address the long time frames necessary for instrument development in oceanography.** Mission agencies, whose future success will depend increasingly on instrumentation that does not yet exist, should initiate suitable roles in the development of new technology.

Transfer of Responsibility

The division of tasks between academic scientists and agencies will depend on the agencies' missions, resources, and internal capabilities vis-à-vis the academic community's. Mechanisms must be developed to provide smooth transition from research activities to operational measurements. In particular, the proposed global ocean observation system will necessitate unprecedented levels of monitoring. **The board recommends that academia and federal agencies work together to ensure that appropriate long-term measurements are extended beyond the work of any individual scientist or group of scientists and that the quality of such measurements is maintained.**

Data Management and Exchange

The board recommends that the present system for data management and exchange within and among the various elements of the marine science community be modernized to reflect the existence of distributed computing systems, national and international data networks, improved satellite data links, and on-line distribution of oceanographic data. Also, provision must be made for future access to existing data.

SPECIFIC PARTNERSHIPS

These general recommendations form the basis for building new partnerships between federal and academic interests in ocean science. Of course, they do not apply to all agencies to the same degree. This section discusses aspects of specific partnerships of the academic oceanography community with each federal agency having a significant ocean program.

Oceanography is now supported by a number of federal agencies using a variety of mechanisms. Federal-academic arrangements differ; the paternal care by the early ONR immediately after World War II, the creation of NSF to foster basic research, the mandated joint fiscal partnership of the National Sea Grant College Program, and cooperative agreements between academic institutions and federal laboratories are salient examples. This section explores aspects of establishing new partnerships between academia and several federal agencies: NSF, the Navy, NOAA, EPA, the Minerals Management Service (MMS), the National Aeronautics and Space Administration (NASA), DOE, and the U.S. Geological Survey (USGS). The discussions are not meant to be inclusive. Further, these discussions are sketches of issues and possibilities, not definitive blueprints. The design of new partnerships and their sustenance must be a fully collaborative process between agency representatives and marine scientists in academic institutions. Some collaboration has already occurred; other cooperative arrangements need to be developed.

Partnerships between the academic community and the agencies that fund ocean research can be improved in several ways. One major improvement would be for the academic institutions to make it career enhancing and attractive for scientists to serve as short-term scientific officers (rotators) at federal agencies. There is a perennial shortage of rotators at these agencies. Rotators should be respected among their peers within the academic community, and assignments should be chosen carefully to benefit both the government and the scientist. Also, scientists should be rewarded for service on federal advisory panels and on community-wide management groups such as the committees of the Ocean Drilling Program.

National Science Foundation

The National Science Foundation was formed in 1950 to increase the nation's base of scientific and engineering knowledge and to strengthen its ability in research and education in all areas of science and engineering. NSF supports fundamental, long-term, merit-selected research in all the scientific and engineering disciplines, including oceanography. NSF maintains strong relationships with academic scientists and is the major source of funding for basic ocean research.

NSF depends heavily on external scientists for program management, program review, individual peer review of proposals, and review panel memberships. The Division of Ocean Sciences (OCE) is the primary supporter of ocean science research within NSF, with specific programs for physical oceanography, chemical oceanography, biological oceanography, marine geology and geophysics, ocean technology, the Ocean Drilling Program, and a program to support facilities for oceanography. Ocean science research is also supported by the Division of Polar Programs, Division of Atmospheric Sciences, Division of Earth Sciences, and Division of Environmental Biology.

OCE depends on its Advisory Committee on Ocean Sciences (ACOS), which prepares long-range plans for the Division of Ocean Sciences. These plans, prepared with input from the ocean science community, identify needs and priorities for ocean science research and research infrastructure. The past two plans were reviewed by the Ocean Studies Board (OSB). A new strategic plan for ocean sciences is being prepared by ACOS, and OSB is expected to be involved.

The Ocean Studies Board (in conjunction with the NRC Board on Earth Sciences and Resources) reviewed the Ocean Drilling Program Long-Range Plan. NSF also depends on outside groups for program and facility management. For example, the Ocean Drilling Program receives advice from the Joint Oceanographic Institutions for Deep Earth Sampling, an international consortium with advisory groups of scientists from the academic community.

The present partnership is basically healthy, and the continued vigor of marine science will depend more than ever on NSF leadership in maintaining the fundamental science. Numerous aspects of the partnership require constant attention: the need for NSF to broker interagency funding for basic science as its own resources are outstripped; the balance between organized scientific efforts and individual investigator, independent grants; and determination of the proper balance among disciplines.

Department of the Navy

The Office of Naval Research has enjoyed a healthy partnership with the academic oceanographic community since its inception. Specifically, ONR funded basic academic research and was largely responsible for the early development and maintenance of oceanography. The academic partnership with ONR has been intellectual as well as financial. ONR depends on external scientists to review its programs through site and program-level reviews and to help develop its science programs through topical workshops. ONR also receives academic advice on program opportunities from the Naval Studies Board and Marine Board of the NRC and the Navy Committee of the Ocean Studies Board. Additional academic input is gained from rotators who come to ONR from the academic community for a few years and then return to academia. ONR's support of academic ship operations has declined in the past few years, which has led to questions about its balance of field and theoretical programs. A joint ONR-academic study of this balance would be useful.

With the end of the Cold War, the focus of Navy-funded research is almost surely going to shift, along with the general level and direction of Defense Department funding. For example, it has been suggested that the recent war in the Persian Gulf implies a greater focus on nearshore problems. However, the Navy, along with NSF, has been the backbone of the U.S. commitment to basic ocean science with a long-term view. Any diminishing of that commitment can, in the long run, undermine both science and national security. The board notes, for example, that the Office of Naval Research is virtually the only federal agency supporting basic research in ocean acoustics.

The Navy recently completed a major consolidation of its laboratories. The result is one "corporate" laboratory, the Naval Research Laboratory (NRL), and four centers: the Naval Air Warfare Center, the Naval Surface Warfare Center, the Naval Undersea Warfare Center, and the Naval Command Control and Ocean Surveillance Center. These organizations, which primarily conduct research on weapon systems and sensors, provide limited general funding and program support to the academic research community. In addition, NRL has a strong continuing relationship with the applied physics laboratories of four universities: Johns Hopkins University, the University of Washington, Pennsylvania State University, and the University of Texas at Austin. As the nation faces budgetary constraints, it is likely that NRL and its centers will explore more cooperative activities with the academic research community, especially in light of the reduction in number of the Navy's dedicated oceanographic ships.

The Office of Naval Technology supports Navy laboratories, universities, and private corporations to carry out its mission in the Navy's Exploratory Development (6.2) program. The academic institutions refine and transfer basic research results into technical feasibility and demonstration plans.

The Oceanographer of the Navy, who serves on the staff of the Chief of Naval Operations, is primarily responsible for providing the oceanographic products and services needed by the Navy's operational forces. In terms of direct funding of research, the Office of the Oceanographer of the Navy and its supporting organizations have only a modest relationship with the academic research community. However, the oceanographer's office provides the oceanographic community with access to global data sets and modeling capability. Data available from the Navy's monitoring network could be an important component of a global ocean observing system. The Navy possesses classified data about the ocean that could benefit ocean science research without compromising national security. It is noteworthy that the Office of the Oceanographer of the Navy has worked over the past three years to declassify much of the data it possesses on seafloor and sea surface topography. Oceanographers look forward to receiving access to more of the data possessed by the Navy. Also, the

Oceanographer of the Navy sponsors all its new oceanographic ship construction, including Navy-owned research ships that are operated by academic institutions. As part of the modernization of the Navy's 1960-vintage oceanographic fleet, the Oceanographer of the Navy ordered three new ships (AGOR class) for the academic research community. The first of these 275-foot-long, multipurpose, deep-ocean-capable research ships (R/V *Thomas Thompson*) was delivered in 1991 to the University of Washington. One of the remaining two new ships will be operated by Scripps Institution of Oceanography and the other by Woods Hole Oceanographic Institution.

An important initiative begun by the Oceanographer of the Navy in 1990 was the sponsorship, in cooperation with the Chief of Naval Research and the OSB, of a tactical oceanography symposium to familiarize the academic community with the Navy's operational needs and requirements. This initiative has become an annual event, and the Office of Naval Technology joined as one of the sponsoring organizations in 1992. The Oceanographer of the Navy is striving to facilitate closer links between the operational side of the Navy and the research community.

National Oceanic and Atmospheric Administration

The National Oceanic and Atmospheric Administration was formed in 1970 from a combination of existing government entities. Its mission is to explore, map, and chart the global ocean and its living resources and to manage, use, and conserve those resources; to describe, monitor, and predict conditions in the atmosphere, ocean, Sun, and space environment; to issue warnings against impending destructive natural events; to assess the consequences of inadvertent environmental modification over several scales of time; and to manage and disseminate long-term environmental information.

Several partnerships now exist between NOAA and the academic community. The National Sea Grant College Program provides support for the study of estuaries and coastal regions, marine applied research, and the application of research to practical problems. Sea Grant is different from most other government-funded research programs in that it is a mandated partnership. Every two dollars of federal funds must be matched by at least one dollar, often from state agencies. Because of this mandated fiscal partnership, policy makers at the state level are generally more aware of Sea Grant research than of research sponsored by other federal agencies. Sea Grant also provides public service through its marine extension and public information components. The partnership has been successful, but its prospects for growth are limited by budget constraints. Further interaction with the academic community will depend on whether the program can find new directions that will justify increased funding.

NOAA also has marine laboratories located near academic oceanographic institutions. The laboratories often support graduate students who carry out thesis research of direct interest to NOAA. Educational opportunities for federal employees range from formal degree programs to seminars and library facilities available at universities. There are also opportunities for cooperative research programs and several NOAA/university joint institutes have been developed. Cooperative agreements between academic and federal laboratories should be expanded to develop stronger intellectual ties between NOAA and the universities.

NOAA provides modest extramural research funds as part of its Climate and Global Change Program and its Coastal Ocean Program, and through the National Marine Fisheries Service. These programs are a good start, showing agency recognition of the need for a broad base of support. The extramural programs should be strengthened to lend stability and to develop close intellectual ties, which are essential if the research is to meet agency needs. Further, critical issues such as the transition of the global ocean observing system to an operational phase must be examined in the context of NOAA's overall responsibilities and of research results from the Climate and Global Change Program and the Coastal Ocean Program. Development and implementation of a global ocean observing system, led by NOAA, would require better partnerships among agencies and between NOAA and academic scientists.

NOAA and the academic community should together evaluate the effectiveness of NOAA research. This examination should include the existing NOAA/university joint institutes, environmental research laboratories, and extramural research support. Improving the quality of scientific research within NOAA, clarifying its role vis-à-vis external science and agency missions, and stabilizing support for extramural research over the long term are clearly in the national interest. A panel of outside experts should work with NOAA's administration to review alternative approaches for extramural and intramural support, including the merits of different funding mechanisms (e.g., the NSF peer-review model, the ONR omnibus contract model, and the existing NOAA cooperative institute model).

The present NOAA fleet consists of 23 ships, of which 5 are inactive and many are old compared to the UNOLS fleet. NOAA's fleet is used primarily to carry out its operational mission in mapping, charting, and fisheries assessment, as well as NOAA research. The fleet occasionally supports other federal and state agencies, academic institutions, and private industry through various arrangements. For several years, NOAA has experienced funding shortfalls for ship operations, resulting in unmet program requirements. NOAA will have to replace its aged fleet and/or use ships owned by others. Under a cooperative arrangement with the academic community, NOAA Corps officers operate the *Vickers,* owned by the University of Southern California. This experiment has not yet concluded and thus has not been evaluated. NOAA and the academic institutions should consider other mechanisms for cooperative ship use, including the use of academic ships by NOAA scientists.

Discussion of the future shape and use of NOAA research vessels should take place within the larger debate on how to manage, upgrade, and use the research vessels operated by all agencies. The concept of a national research fleet is providing a context for this discussion. It is clear that we can no longer afford the luxury of regarding individual agency vessels as unrelated, with no sharing of resources.

A major obstacle for marine science lies in the difficulties of developing and managing spaceborne instruments over the next decades. Historically, NASA developed meteorological spacecraft that eventually evolved into operational systems managed by NOAA. However, for marine observations, apart from long-standing efforts in visible and infrared sea surface temperature observations and microwave sea ice measurements (both of interest to short-term forecasting), there is no effective mechanism for the systematic development or transfer of technology from research to operations. Some mechanism must be found to routinely collect such observations that are important to the NOAA mission. NOAA will need additional funding to carry out these observations, and a partnership arrangement will be necessary to identify the essential variables to be observed.

Another area of potential partnerships involves data bases, especially their accessibility. NOAA is responsible for the National Oceanographic Data Center (NODC). Created in the 1950s, NODC is intended to provide both present access to data and an archive for future generations. However, the center has failed to keep abreast of changing technologies in observation and data base management. As global programs generate increasing volumes of data and place new demands for the use of data from all sources, the need for modern national data facilities will become increasingly urgent. Because working scientists are often the source of many of the data and are often the largest potential users, they should participate in the design and use of these important data bases. The Joint Environmental Data Analysis center at Scripps Institution of Oceanography, which involves active scientists in the quality control and decisions of archiving data, is a first step in developing such partnerships.

Environmental Protection Agency

Since its founding in 1970, the Environmental Protection Agency has developed numerous regulations relative to both air and water, and environmental quality in many previously heavily polluted areas has improved as a result of these controls. Now, as environmental problems on regional, national, and international scales are increasingly recognized, EPA's challenge is to improve our understanding and management of the sources of pollutants and the environments that receive waste. The EPA Science Advisory Board (1990), in its landmark report *Reducing Risk*, stated that too little attention is paid to environmental problems that have significant large-scale consequences and low reversibility (e.g., global climate change and loss of habitats and biodiversity). In the past, EPA has relied on internal expertise for scientific input, but the range of problems and their complexity can no longer be handled in this way. EPA has made a commitment to the increasing use of scientific advice throughout its activities. Meeting this commitment will require strong partnerships with the academic community.

EPA's need both to view pollution control from a larger environmental perspective and to increase its reliance on science offers prospects for partnerships with the academic ocean science community. EPA engages scientists in its environmental research laboratories, a relatively small extramural grants program, exploratory environmental research centers, and environmental management programs, including the National Estuary Program. An expanded EPA partnership with the academic community could include the following:

• expansion of the extramural grants program and creation of additional environmental research centers collocated with universities and specifically focused on present and future problems in the marine environment;

• agreements between EPA research laboratories and nearby academic or private institutions;

• training of EPA personnel in newly emerging science that enhances the science perspective in order to balance the strong regulatory perspective that exists within the agency; and

• increased reliance on academic experts in areas in which they may be better positioned than commercial consultants (e.g., analysis of long-term and large-scale environmental problems).

Problems with the agency's approach to academic grants and centers have discouraged many university-based experts from working with EPA. In addition, the program and regional offices and the Office of Research and Development laboratories often rely on contractual mechanisms that prevent EPA from obtaining the best outside scientists to work on agency issues. EPA should move quickly to bolster its grants and centers programs. The agency should also implement a long-term plan to replace contractual mechanisms that may be detrimental to obtaining the best possible scientific information.

Minerals Management Service

In 1973, in response to the threat of an international oil embargo, President Nixon announced an ambitious program for accelerated exploration and development of the oil and gas resources of the outer continental shelf (OCS) of most of the United States. Although it had managed offshore development in the northwestern Gulf of Mexico for many years, the Bureau of Land Management (BLM) was suddenly required to evaluate the environmental consequences of greatly expanded exploration and development. Since that time, through the BLM (now MMS) Environmental Studies Program, the Department of the Interior has spent more than $259 million for studies of the climate, circulation, contaminant levels, ecology, living resources, geohazards, and effects of oil and gas development in all OCS areas, particularly those with no previous development.

Although many academic ocean scientists have been involved in MMS studies, the agency has traditionally relied on commercial procurement contracting to acquire technical information. Some consequences of this approach are that relatively little of the information produced was published in the open scientific literature, whereby it could undergo peer review and perhaps gain broad credibility, and a cadre of environmental scientists, who could influence public opinion and policy, was not nurtured. Further, the program's emphasis on short-term results as opposed to long-term understanding provided limited opportunity for research innovation. To overcome these limitations, MMS has sought to increase the involvement of academic ocean scientists in its Environmental Studies Program through a variety of mechanisms: (1) two cooperative agreements with university groups to support investigator-initiated research on the long-term effects of petroleum development activities (i.e., the Louisiana Universities Marine Consortium in the Gulf of Mexico and the University of California-Santa Barbara in southern California); (2) other cooperative agreements with academic institutions that have unique capabilities for meeting MMS information needs; (3) the award of competitive contracts for large projects to academic institutions (e.g., Louisiana-Texas shelf physical oceanography studies at Texas A&M's Texas Institute of Oceanography and Louisiana State University); (4) extensive involvement of academic oceanographers on the scientific committee of the OCS advisory board and on quality review boards of various studies; and (5) increased emphasis on publication of study results in the open scientific literature.

MMS is already actively seeking to develop partnerships with academic oceanography, but to further these relationships, it should consider the following:

• expansion of the cooperative agreements for strategic investigator-initiated research on long-term environmental and socioeconomic effects of oil and gas in developed OCS regions;

• use of academic institutions (similar to the recently initiated physical oceanographic studies in the Gulf of Mexico and off California) for complex scientific studies that require the innovation and integration for which these institutions are particularly well qualified; and

• participation in the shared use of the academic research fleet with other federal agencies through more active involvement with UNOLS. MMS research vessel requirements and scheduling constraints do not always coincide with the availability of UNOLS vessels.

National Aeronautics and Space Administration

The National Aeronautics and Space Administration develops new technology for space, demonstrates its use for a variety of scientific and technical purposes, and supports related science. NASA-developed technology provided the first synoptic views of Earth, and NASA Earth observation programs have since evolved into the present international operational and research missions for remote sensing of processes in the atmosphere and at the ocean and land surfaces. The great difficulty in observing the ocean by conventional means (ships and buoys) led oceanographers early in the post-*Sputnik* period to recognize the value of spaceborne observations.

In the more than 30 years since satellite imagery was first demonstrated, NASA and the ocean community have achieved notable successes. Satellite-measured sea surface temperatures are now routine input for weather and climate forecasting. NASA guided this technology to its present mature operational state. The *Seasat* and *Nimbus-7* missions demonstrated the validity of the idea that the ocean surface's shape and color could be measured from space and would be useful. Data from these two satellite missions are still used by ocean scientists.

As part of the Earth Observing System (EOS), NASA plans a major data and information system, the Earth Observing System Data and Information System (EOSDIS). EOSDIS will contribute to the Global Change Data and Information System, a joint venture of NASA, NOAA, and USGS mentioned earlier. Oceanographic data will form an important part of these data systems, and the oceanographic community should ensure that it is well represented on the advisory and management groups for these systems. Beginning in the early 1980s, NASA worked with the academic oceanography community to develop a plan for satellite oceanography and to build a first-class national oceanographic satellite capability. NASA established excellent scientific centers at the Jet Propulsion Laboratory and the Goddard Space Flight Center, and put together an effective headquarters team that oversaw the centers' research and supported research at academic institutions, many of them outside the mainstream oceanographic institutions. This effort, which was endorsed at the highest levels of the agency, led to a period of extremely effective collaboration and joint projects. Both NASA and the institutions learned from each other: NASA, a large federal agency oriented toward massive team efforts extending over many years, and the research community, which is often interested in smaller projects lasting no longer than a graduate student's thesis period.

The investment that NASA made in marine science in the

1980s is about to pay off in a surge of data from missions using satellites that will fly in the 1990s. Considerable expertise and experience now exist both within the NASA centers and in the nonfederal laboratories and universities—almost all of which can be attributed to the far-sighted NASA policies of a decade ago. The only parameter strongly recommended by the ocean community for measurement in the 1990s that is not included in present plans is Earth's gravity field; this oversight needs to be rectified by joint discussions between NASA and the European Space Agency.

As we look beyond the 1990s and well into the twenty-first century, a favorable outlook is not so clear for ocean satellite measurements. In the past several years, NASA has focused primarily on EOS, a series of satellites aimed at contributing to global change research. EOS's task is to provide a wide variety of data in the late 1990s, but limited budgets are reducing the number of instruments and delaying the launch of others. Certain segments of the ocean community have been involved in EOS planning, but the connection is not as broad as it should be. Moreover, the oceans branch at NASA headquarters has been subsumed into EOS planning, thus eliminating the focal point for ocean interests within NASA.

With this lack of focus, it is more difficult for ocean science to be heard regarding ocean priorities in space measurements. As a result of recent EOS downsizing, ocean instruments have lower priority, and the missions needed for broad coverage of ocean parameters in the twenty-first century are not well defined. If long-term planning does not begin soon, the required missions will not be available to provide continuity with missions flying in the 1990s.

Another problem is alluded to in the discussion of NOAA. For climate purposes, long continuous time series of ocean measurements must be sustained. Because of the requirement for open-ended measurements, the measurements resemble operational ones. Traditionally, NASA has asserted that it did not make operational measurements—that the technology would be transferred to NOAA for that purpose; but NOAA has not received adequate funding even for the limited measurements to be made from the polar and geostationary operational environmental satellites. A closer connection is needed between NASA and NOAA in the transition from research to operations. This problem has been identified by several national advisory committees; it was brought to the attention of the responsible interagency committee, the National Space Council, and is being debated there. Because global change research is a national concern, resolution of this problem of transition is urgent.

The transition of NASA technology to Department of Defense (DOD) operational measurements has had mixed success; the microwave radiometer is now operational in the Defense Meteorological Satellite Program, and data are provided routinely to academic investigators. The Navy has flown, and plans to continue to fly, additional altimeters for ocean surface measurements. However, NASA's attempts to work with DOD on the flight of other instruments for surface winds and ocean color have floundered; this area also needs attention.

Because of the importance of oceanography to the Global Change Research Program, NASA should reestablish some mechanism with sufficient stature at headquarters to communicate with the marine science community. NASA should formulate, in collaboration with other agencies and the academic community, a coherent sense of where its long-term responsibilities lie for the overall health of marine science. For example, NASA is the agency that can nurture the special scientific and technical expertise required for the use of satellite remote-sensed data, and it must do so. Partnerships are key; it is more important than ever for the ocean community to develop partnerships with NASA, as it has with other agencies. NASA should help foster these partnerships. Further, NASA needs to recognize the importance of supporting surface-based programs that both directly support and help maximize the scientific returns from its spacecraft.

Department of Energy

The Department of Energy, formed in 1977, is responsible for supporting the development of energy production and conservation technology, the marketing of federal energy supplies, nuclear weapons research and development, energy regulation, and the collection and analysis of data on energy production and use. DOE has carried out marine-related research for many years, most recently as part of its Carbon Dioxide and Coastal Ocean Margins Programs in the Office of Health and Environmental Research. The research focused initially on understanding the fate of radionuclides. DOE marine research is presently concentrated on chemical and biological aspects of the global carbon cycle to understand the fate of the carbon dioxide emitted to the atmosphere as part of energy production and use. In particular, DOE has funded studies of integrated regional biological productivity on the continental shelf, the cycling and transport of organic carbon and nutrients across the shelf, the influence of western boundary currents (e.g., the Gulf Stream) on shelf physics and biological productivity, particle transport processes, and particle burial in basins along the continental margin.

DOE is one of the few agencies to support long-term research in coastal oceanography. Long time series are useful to determine whether the coastal ocean is changing because of anthropogenic influences and to separate directional changes from natural variations. Earlier programs supported the development of in situ instruments to measure optical properties, particle concentration and flux, chlorophyll, and nutrients, allowing important scientific advances. DOE's support of the successful Food Chain Research Group at Scripps Institution of Oceanography is an example of the value of its early academic partnerships.

Somewhat more than a decade ago, Congress assigned DOE the responsibility to collect information and maintain a major data base on carbon dioxide. Interest in carbon dioxide was growing because of the increasing body of theory suggesting a relationship between the greenhouse effect and energy production and supply. As part of the interagency focus on global change research, several programs initiated within DOE in the past few years capitalize on its experience and interests. Two major programs have emerged: the Atmospheric Radiation Measurements (ARM) program and the Computer Hardware, Advanced Modeling and Model Physics (CHAMMP) program. The ARM program is designed to make complete and detailed measurements at strategically chosen sites to enhance our understanding of clouds and solar radiation. The primary focus of CHAMMP is climate modeling. One of its major goals is to advance the speed of climate models by using highly parallel new computer hardware systems, other software techniques, and new algorithms. Many of the major ocean-atmosphere models from around the world are now being compared. In addition, DOE is requesting an increase in the fiscal year 1993 budget for its open ocean research thrust to fulfill its mission to understand the carbon dioxide balance and the ocean's role in this balance.

DOE funds both extramural research and research carried out at its national laboratories. DOE's national laboratory system employs approximately 50,000 people and has a budget of $6 billion to $8 billion. Marine research is a small part of the overall DOE research effort; Brookhaven National Laboratory is the primary site for marine research. As the oceanographic community discusses its approach to the partnership it hopes to forge with DOE, the laboratories should be considered integral participants.

DOE has sought scientific advice on its marine research through workshops, standing committees of the National Research Council, and one-time reviews by NRC panels and other groups. The OSB has reviewed the Coastal Ocean Margins Program, has advised the Carbon Dioxide Program on oceanic carbon dioxide research, and is presently advising the Office of Health and Environmental Research on the application of molecular biological techniques to marine research. The Coastal Ocean Margins Program would benefit from a standing panel of outside experts to help its staff formulate a focused research plan that would build on the agency's strengths in long-term monitoring and regional research. The existing DOE partnerships with academic scientists in the Carbon Dioxide Program and in the area of molecular marine biology appear stronger.

A more general issue for the oceanographic community to consider is where, in light of DOE's missions, new common grounds might lie. There appears to be a genuine interest on DOE's part to enhance or change its role vis-à-vis Earth sciences. The energy implications of marine geology and geophysics research seems to be a natural field for initial discussions. The plans now being developed by DOE for small satellite missions to measure radiation might well be enhanced to include small satellite missions for ocean measurements. There are clearly many other areas in which energy research and energy supply options overlap with ocean science interests. Future partnership discussions with DOE might be aimed at assessing priorities and planning possible interactions in each particular area.

U.S. Geological Survey

The U.S. Geological Survey was established in 1879. Its primary responsibilities are identifying and characterizing the nation's onshore and offshore land, water, energy, and mineral resources; investigating natural hazards (e.g., earthquakes, volcanoes, and landslides); and conducting the National Mapping Program. To achieve these objectives, USGS prepares maps and digital and cartographic data; collects and interprets data on energy, mineral, and water resources; performs fundamental and applied research in Earth sciences to understand Earth processes and their variations in time and space; and publishes and disseminates the results of its investigations in the form of maps, data bases, and reports. The USGS marine program has two components: (1) the Offshore Geologic Framework, and (2) Coastal and Wetlands Processes. The Offshore Geologic Framework components conduct regional scientific investigations aimed at understanding and describing the geologic framework, energy and mineral resources, geohazards, and seafloor environmental conditions of U.S. offshore and other areas that could potentially provide a continued supply of needed resources.

The overall objective of USGS coastal research is to improve our ability to predict coastal erosion, wetland loss, coastal pollution, and the location of marine hard mineral resources through a better understanding of processes and the geologic framework within which the processes operate. Improved predictive capabilities are needed by coastal zone planners and managers and are required for preservation of the nation's coastal resources. Thus USGS marine science activities range from a major systematic mapping of the U.S. Exclusive Economic Zone (EEZ), to deep seismic exploration beneath the seafloor and continental margins, to transport processes within the ocean and in coastal areas. Recent increased focus on the coastal zone resulted from government interest in sea-level rise and pollution. Because USGS participates in many national and international research programs with academic scientists, it has developed effective means for peer review and communication of agency research results. An example of partnerships is the USGS Marine Program, begun in the 1960s. The program located it facilities near academic or oceanographic institutions (i.e., Woods Hole Oceanographic Institution, the University of South Florida, Stanford University, and the University of Washington), which permits sharing of marine infrastructure and human resources. Numerous memoranda of understanding and cooperative agreements with other universities are also in place for specific program tasks and needs. USGS annually conducts a part of its field operations on UNOLS ships.

An expanded partnership between USGS and academia could include the following:

• Increased use of external scientists to review the USGS ocean science program. This process might help to clarify the unique role of USGS in marine research. Aspects of the USGS Marine Program are presently reviewed by the Marine Board and other NRC boards.

• Increased participation of external scientists on collaborative projects. Examples of recent successes include studies conducted in California, Boston harbor, and Louisiana and the participation of academic scientists and students in EEZ mapping cruises.

• Reexamination of USGS marine research goals in light of areas for increasing cooperation with academic scientists.

3

Future Directions in Ocean Sciences

The possibility of and the need for studying the ocean on a global scale provide a major impetus for new partnerships in oceanography. The design and deployment of a global ocean observing system, now being discussed, will be possible only with the cooperation of ocean scientists and governments throughout the world.

THE SCIENCE OF OCEANOGRAPHY

Oceanography, the science of the sea, serves many purposes while deriving impetus from many sources. All of oceanography—physical, chemical, geological, and biological—is driven by scientists interested in advancing basic knowledge. Ocean scientists have made a number of exciting discoveries in the past 30 years that have changed our view of Earth. The discovery of oceanic eddies has been important for an understanding of ocean circulation, propagation of sound in the ocean, fisheries productivity, and other ocean processes. Verification by ocean drilling that Earth's crust is divided into moving plates that are created at mid-ocean ridges and recycled into Earth's interior replaced the traditional view that the surface was essentially static. Discovery of dense colonies of animals and bacteria at some deep-sea hydrothermal vents demonstrated that organisms could thrive in ecosystems based on chemical energy from Earth's interior rather than directly on energy from the Sun. Study of the combined ocean-atmosphere system has provided sufficient knowledge of interannual climate variations that scientists are now able to forecast El Niño climate disturbances months in advance.

Over the next decade, oceanography will continue to provide exciting discoveries by contributing new understanding of Earth as a system and by helping us understand how humankind is altering the system. It is now essential (and possible) to study ocean processes on a global scale. The oceanography of the next decade will take place in the traditional marine science disciplines and at the boundaries of these disciplines. New partnerships among oceanographers working in different disciplines should lead to new discoveries about the ocean's role in climate change, the function of mid-ocean ridges, and coastal ocean processes.

Additional oceanographic studies in the coming decade will focus on how ecosystems affect global cycles of important chemicals and, conversely, how changes in the global environment affect marine ecosystems. Studies of ecosystems at hydrothermal vents and hydrocarbon seeps will refine our ideas about the conditions under which life is possible and about the origins of life. More of the ocean floor must be explored to determine the extent and nature of deep-ocean vents, their ability to support novel organisms, and their importance in global chemical cycles. Continued study of the ocean's chemistry should bring new understanding of the past state of Earth, how ocean processes operate today, and the contribution of sources and sinks of various chemicals. The study of deep-ocean sediment cores will provide more information about past natural cycles of Earth's climate, with which present climate fluctuations can be compared. Oceanographers will achieve a better understanding of the variability of the circulation of the world ocean. The interaction of climate with this circulation is only poorly known, but there is evidence that the transport of surface water to depth can vary greatly even over as short a time as one decade.

Unlike many other sciences driven by scientific curiosity, aspects of marine science have immediate and obvious practical applications. These include, but are not limited to, the control of climate by ocean circulation, chemical and biological reactions to climate change, understanding fisheries productivity, movement of pollutants, and the problem of coastal development in the face of rising sea level. Oceanographers are fortunate to take part in a science that is fascinating, compelling, and intellectually challenging. Oceanography is also a science whose outcome is of immediate societal application and in which the financial stakes are potentially immense, for example, the economic impact of a reliable forecast of a sea-level rise. Because the societal implications of the science are readily apparent to policy makers, they may demand answers to purely practical questions in the short term. This pressure can distort the investment in basic science, undermining the quest for basic understanding that remains key to the long-term solution of practical problems. Thus the functioning of oceanography in the United States should focus both on sustenance of the underlying basic science and on specific answers to practical questions of short-term urgency.

This chapter summarizes the concerns of basic scientists, with some focus on the interaction of basic science with more practical problems. Several themes are common throughout the discussion, which is divided by classical disciplines. First is the growing sense that the basic science now encompasses the global ocean scale. This capability and the need to conduct global-scale studies have led to the planning of large-scale, long-term cooperative experiments. Primarily planned and executed with National Science Foundation (NSF) support, they focus the work of many scientists on global ocean research. These large programs are usually managed through national or international consortia that involve many scientists, agencies, and often countries. Such programs will explore new questions and test new mechanisms for working together in the next decade. Global uncertainties are rapidly moving much of oceanography from the capabilities and interests of single or small groups of investigators for a limited time to the involvement of many individuals, institutions, and governments for decades. Mechanisms must be developed for these new large-scale efforts to be sustained in a scientifically and technically sound manner, by coordinating the plans of other nations, federal agencies, academic institutions, and individual scientists.

Second, all sections of this chapter emphasize the dependence of the subject as a whole continued technical developments. The ocean is remarkably difficult to study, given its size, opacity to electromagnetic waves, and general hostility (e.g., its corrosiveness, high pressures, and turbulence). The health of all disciplines depends directly on the continued development of new tools designed to solve their fundamental sampling problems. In the past decade, oceanographic sampling improved through incorporation of new technologies from other fields, such as remote sensing, material science, electronics, and computer science. A fundamental change arising from the use of these new technologies is an increase in the quality and the volume of data collected. Accompanying this change is a significant increase in each oceanographer's capacity to study ocean phenomena, an increase that raises the costs for each oceanographer's science. As the cost per oceanographer for scientific equipment and facilities has increased, the field has responded with increased sharing of facilities, such as ships and submersibles, and equipment, such as the new accelerator mass spectrometer for carbon-14 measurements. The development and shared use of expensive facilities are likely to continue in the future. Yet even with shared facilities, inflation-adjusted research funding for the ocean sciences has remained nearly constant over the past decade, while the number of Ph.D.-level academic oceanographers has increased by about 50 percent and societal pressures to predict man's effect on the ocean have also increased. The growth in the scientific capacity of each investigator and the number of qualified investigators, coupled with nearly constant funding, has resulted in partial funding for some ocean researchers.

Third, the resolution requirements of oceanographic models and the complexity of model physics have always outstripped the largest computational capability anywhere. As understanding of the ocean becomes more sophisticated, more sophisticated models are required. The nurture of computational capability is reflected across the disciplines.

Fourth, the understanding of the ocean and of the problems of oceanographers has progressed so much in the past several decades that all disciplines are now capable of new accomplishments in a seemingly endless number of areas. The problem is that the potential far exceeds the resources likely to be available, and the difficult task of setting priorities within and across disciplines will be amplified.

The foundation of knowledge about the ocean that is now used in policy decisions was gained largely through Office of Naval Research (ONR) and NSF investments in basic research over the past several decades. Yet the demand for quick answers to purely practical questions sometimes obscures the need for investing in basic science, which remains the key to long-term practical applications. Under pressure to provide immediate solutions, mission agencies may be tempted to focus only on the short term. One example of the importance of basic research is a 1961 study that is now contributing to the debate about climate change—the question of whether ocean circulation has two stable states. Both the geological record and numerical models suggest that, at some times in the past, ocean circulation was unlike today's and that it could switch rapidly from its present state to a radically different one. When the ocean was in this alternate state, Earth's climate was not at all like today's. This idea dates back to a paper written by Henry Stommel (1961), an academic scientist driven primarily by his own curiosity and supported by ONR and NSF. The paper had little impact for more than 20 years. Now regarded as seminal, it illustrates the need to sustain basic science so that future generations will have a knowledge base from which to develop their policy decisions.

The authors of the following sections were asked to discuss the dominant issues of their disciplines and to lay out the grand themes, providing a scientific underpinning to discussion of the new partnerships. Ten years is probably the outer limit of an attempt to suggest what the major science themes will be. A decadal report written in 1960 would almost surely have missed the revolution in plate tectonics and thus would have been hopelessly wrong in its discussion of some dominant scientific themes in 1970. On the other hand, such a report could have captured accurately the methodologies of work at sea and the human resource requirements. Of course, the central questions of the field did not change either—although an intellectual revolution in the way they could be discussed occurred.

The decision to organize this chapter according to traditional oceanographic disciplines was not arbitrary (the coastal ocean is a special case, discussed below). Anyone who reads each section will perceive exciting and important scientific problems that cut across many or even all disciplines. Examples are the growing importance of paleoceanographic studies that involve geology, geophysics, chemistry, biology, and physical oceanography because of their climate implications. Likewise, the study of ridge crests cuts across geology and geophysics, biology and chemistry, and even slightly, physical oceanography. Nonetheless, the board believes that there is a danger in declaring such interdisciplinary studies as the likely focus of future marine science efforts. Without denigrating the science done on such problems, interdisciplinary studies clearly build on the foundations of chemistry, physics, geology, geophysics, and biology. These, in turn, depend directly on their nonmarine counterparts of physics, mathematics, numerical methods, and other fields that provide the intellectual fertilization of marine studies. The history of ocean sciences suggests that one cannot have good interdisciplinary science without good disciplinary foundations, and it is essential that the traditional oceanographic disciplines retain their identity and vitality. One needs to encourage scientists working on interdisciplinary problems, but they must first be expert in one or more of the basic disciplines. Just how such fostering should take place is the subject of debate, and the reader will detect a degree of disagreement as to how we should move forward. Working out various combinations of scientists and institutions is a major challenge for our academic institutions in the next decade. The board makes no specific recommendation except to note that the strength of the U.S. scientific community is its ability to tolerate and encourage great diversity in its institutions.

Because the following sections were written by a number of different authors, they differ in style and content. The sections are not meant to be all inclusive but instead to provide a flavor of the excitement of each discipline of oceanography.

The treatment here of coastal oceanography is anomalous because it deals with a geographic region—that is, shorelines, estuaries, bays, and the continental shelf—and not a discipline. The large percentage of the U.S. and world population that lives in the coastal zone, and the multiple human uses and impacts on the coastal ocean, place this area of oceanography much more conspicuously and immediately in the public policy arena. Unlike the participants in deep-water marine science, states, cities, and private enterprise are all prominent players in understanding and using the coastal ocean. The interplay of the basic sciences of fluid flow, chemistry, biology, shoreline physics, and geology with public policy concerns leads to a near-term urgency that cuts across scientific disciplines. However, it is important to recognize, as this report does, that the foundations of understanding must rest firmly on the underlying basic sciences.

DIRECTIONS FOR PHYSICAL OCEANOGRAPHY

Summary

The great volume of water in the ocean exerts a powerful influence on the Earth's climate by absorbing, storing, transporting, and releasing heat, water, and trace gases. The goal of physical oceanography is to develop a quantitative understanding of the ocean's physical processes, including circulation, mixing, waves, and fluxes of energy, momentum, and chemical substances within the ocean and across its boundaries. Addressing such problems will require sustained large-scale observations of the world ocean aided by advances in measurement and computational technology. Designing and deploying a global ocean observing system are among the most important and difficult tasks for physical oceanography and climate studies for the next decade. Such a system would incorporate existing measurement programs as well as observations that are not yet routine.

Several topics will dominate physical oceanographic research in the coming decade. Research in modeling, ocean mixing, thermohaline circulation, and water mass formation processes will be important. To achieve their scientific objectives and to make more complete ocean observations, physical oceanographers must use both proven methods and new technologies, including acoustic techniques; measurements made from volunteer observing ships; satellite observations and data relay; and measurements of the distributions of trace chemicals.

Introduction

The ocean consists of nearly 1.4 billion cubic kilometers of salty water, about 97 percent of the free water on Earth. In comparison, the atmosphere holds only about 0.001 percent. This volume of water exerts a powerful influence on Earth's climate by transporting heat, water, and other climate-relevant properties around the globe and by exchanging these properties, as well as greenhouse gases (e.g., carbon dioxide, methane, and chlorofluorocarbons), with the atmosphere. Net ocean absorption of greenhouse gases and some greenhouse-induced heat from the atmosphere can delay greenhouse warming of the atmosphere. Predicting future climate conditions depends on learning what controls ocean circulation and water mass formation, and whether the system is predictable, even in principle.

Physical oceanography, like many fields of science, consists of theory, observations, and numerical models. Physical oceanographic theories use the equations of fluid dynamics, modified to account for Earth's rotation and shape (e.g., O'Brien, 1985). A goal of physical oceanography is to develop a quantitative understanding of the ocean circulation, including fluxes—of energy, momentum, and chemical substances—within the ocean and across its boundaries. Physical oceanographers must contribute to the increasing societal emphasis on measuring, predicting, and planning for changes in global climate by improving understanding of the physical factors that maintain the overall physical, chemical, and biological characteristics of the ocean. Advances in measurement and computational technology will continue to contribute to advances in physical oceanography.

Studies of climate change put the skills of oceanographers to a severe test. The time scales are long: interannual, decadal, and beyond. Physical processes are three dimensional and involve interaction of the ocean with the atmosphere. Winds transfer momentum and promote mixing and evaporation. Atmospheric temperature influences the density of ocean surface layers through effects on seawater temperature and salinity (through ice formation and melting), which in turn modify the atmosphere. Development of the physical state of the ocean is difficult to model because it involves the complex interaction of processes that operate on vastly different time and space scales. Nonetheless, progress is being made. Techniques that will permit better and more frequent observations are being developed, and advances in numerical modeling will soon permit representation of the major components of ocean circulation.

Ocean observations reflect the state of the ocean and hence the forces acting on it. Because observations are made in a corrosive, turbulent environment with high pressures at depth, they are difficult and expensive to obtain. Because of the size and variability of the ocean, measurements are always incomplete in space and time. Yet understanding the ocean depends on adequate measurements, and to make them we need to use technologies that permit a view of the global ocean. Technologies based on acoustics, space-based remote sensing, and underway automatic measurements could all be applied to global-scale observations.

Predictions of the ocean can be carried out only when the initial and boundary conditions are provided from observations with an accuracy and precision consistent with the physics present. Because oceanic observations are so expensive, models and theories must be used to help determine the most cost-effective measurements and measurement systems.

Global Ocean Observing System

Physical oceanographic observations and modeling are becoming global, but the resources required to deploy and sustain large-scale observations of the world ocean are enormous. The exact configuration of a global ocean observing system is unknown, but it would probably include existing observations from satellites, moored open ocean sensors, volunteer observing ships, and the global sea-level network, as well as other observations that are not yet defined or collected routinely. The scientific and techno-logical results from several ongoing large-scale research programs—the Tropical Ocean-Global Atmosphere program, the World Ocean Circulation Experiment (WOCE), and the Joint Global Ocean Flux Study—should be used to design an operational observation sys-tem that is effective, affordable, and consistent with our knowl-edge of the scales of ocean biology, chemistry, and physics. It would be the largest field enterprise ever undertaken by the oceano-graphic community, and it must have an international and multidisciplinary scope well beyond previous experience. The design and implementation of a global ocean observing system (GOOS) must involve ocean scientists substantially because the design is extremely important to the science itself and depends on firm scientific understanding. Designing and deploying a GOOS is one of the most important and difficult tasks for physical ocean-ography and climate studies in the next decade. The United States should take a major leadership role in both the research and the operations.

Because of the present paucity of ocean data, numerical mod-els are important in the development of a GOOS. Models will be used to interpret available data for testing possible system designs and, ultimately, to interpret the data from such a system.

Major Research Topics for the Coming Decade

Several topics will dominate physical oceanographic research in the coming decade. The list is incomplete; the topics men-tioned received some emphasis during the Ocean Studies Board workshops as representing key research issues and include the following: research in modeling; ocean mixing, including interior mixing and the surface mixed layer; thermohaline circulation; and heat and freshwater fluxes.

Ocean Modeling

The central focus of numerical modeling of the ocean has been, and continues to be, directed toward fluid dynamics, but the mod-els have importance far beyond physical oceanography. For ex-ample, communities of organisms in the upper ocean live in a delicate balance, depending on the stability of the water column, its mixing rates, and its large-scale vertical and horizontal fluid movements. Our limited ability to predict the movements of the upper ocean limits understanding of basic biological processes.

The transfer of gases between atmosphere and ocean is central to the carbon cycle; this transfer relies on many scales of circulation and mixing.

Combined atmosphere-ocean simulations at interannual time scales require more accurate ocean models. The present generation of coupled atmosphere-ocean models exhibits unacceptable drifts (Manabe and Stouffer, 1988). Climate forecast models must be free of even small systemic errors that accumulate over long simulated periods, hiding the signals that are sought. To understand and mimic the paleoceanographic record, a major test of global models, one must be able to carry model integrations over time scales corresponding to thousands or tens of thousands of years. It is not clear that ocean and atmosphere behavior is predictable on scales of decades or longer. The limits of predictability are being explored as a research topic.

The global ocean is so large and its circulation occurs over such a variety of space (tens to thousands of kilometers) and time (days to centuries) scales that ocean circulation modeling has always overwhelmed even the largest supercomputers. This situation will probably remain for some decades to come. Thus it is a major intellectual challenge to design models of ocean circulation with time and space increments small enough to model processes adequately, given foreseeable limitations in computing resources.

Prominent features and processes that must be incorporated more accurately into physical oceanographic models (in a manner consistent with observations) include the effects of complex bottom topography on deep-water masses, deep vertical and horizontal mixing, eddies and fronts in the upper ocean, the interaction of water flow and diffusion of a variety of properties, boundary effects at the seafloor and surface, and the dynamics of shallow and deep boundary currents.

Ocean Mixing

Interior Mixing   Large-scale ocean circulation is coupled with, and partially controlled by, small-scale mixing processes. Understanding the places, rates, and mechanisms by which the ocean mixes heat, salt, and momentum is crucial to understanding the circulation of the largest scales and essential to any capability to predict future oceanic states. It is intimidating to realize that to understand the dynamics of large-scale circulation and convective water mass formation, we must also understand the physics acting on the smallest scales (centimeters and millimeters). Heating, cooling, flow, and mixing processes act together to determine physical properties of the ocean. Changes in any one of these processes can affect the global climate system. Significant progress in the observation of ocean mixing processes and in the interpretation of these observations has been made, but understanding remains inadequate.

The capability to compare direct mixing measurements (through microstructure and purposeful tracer releases) with natural mixing (estimated indirectly from natural tracer distributions) is rapidly accelerating our understanding of mixing processes (Watson and Ledwell, 1988). The results of such comparisons will direct future research. For example, if deep-sea observations confirm that mixing rates are lower than predicted, attention will focus on mixing processes in the benthic boundary layer and continental slopes. If tracer studies indicate significantly more mixing than is seen by direct measurement, double diffusive and other mechanisms will be explored. With new observational techniques and a clear measurement strategy, significant progress can be expected in the coming decade in the study of ocean mixing.

Surface Mixed Layer The primary production that supports the entire marine food web occurs in the upper sunlit portion of the ocean (the euphotic zone), where photosynthesis occurs. Our growing concern for climate variation makes understanding the uptake of carbon dioxide related to photosynthesis of particular importance. Fortunately, several developments over the last few years in biological oceanography, marine chemistry, and ocean physics promise advances in the study of the biological-chemical cycles in the euphotic zone. Exploitation of new techniques could significantly improve our ability to predict various aspects of global environmental change, including the ocean's role in sequestering carbon dioxide.

The topmost layer of the ocean is called the "mixed layer" because the waters are mixed by wind, waves, and currents. This layer is often nearly homogeneous in temperature and chemical characteristics, and is bounded by the sea surface and a layer of denser water. The transfer of gases between atmosphere and ocean depends primarily on mixed-layer processes. Understanding the physics of the ocean surface mixed layer, and its coupling with the ocean interior and the atmosphere, is essential if the combined biogeochemical systems of ocean and atmosphere are to be represented correctly in ocean models. Mixed-layer studies are among the endeavors of physical oceanography in which strong interaction with the fields of biology and chemistry is particularly rewarding. The rates of heat, water, gas, and momentum exchange across the ocean-atmosphere interface must be estimated better from easily measured or calculated environmental parameters (e.g., wind, temperature differences across the air-sea interface, waves, and stability of the mixed layer).

Because photosynthesis involves the conversion of carbon dioxide and nutrients into living material and oxygen, the euphotic zone is an immediate sink for atmospheric carbon dioxide. The flow and mixing of water masses and the transport of nutrients and particles that control phytoplankton populations depend on physical oceanographic processes. Understanding the mixed layer is a complex problem involving studies in marine biology, chemistry, and physics. In a crude sense, the mixed layer can be regarded as controlled by a set of chemical reactions in which biological processes determine most of the reaction rates (nutrient fixation and regeneration) and physical processes (advection, mixing, and particle sinking) determine the rates at which reactants and products are provided to or removed from the system.

The physical oceanographer's approach to studying the surface mixed layer involves measurement of currents and horizontal variations to determine advection, microstructure measurements to study turbulent fluxes, and measurements of deeper properties to infer vertical flows. Chemical oceanographers study the latter process using time-dependent tracers to estimate the vertical path of water masses and observe changes along it. Both methods can be strengthened by a model that integrates the measurements.

Thermohaline Circulation   In a few limited regions of the ocean, a combination of low temperature and high salinity produces dense surface water that flows into the deep ocean and spreads laterally to initiate global-scale thermohaline circulation. Deep-reaching convection occurs in the northern North Atlantic Ocean and around Antarctica. These water masses spread throughout the ocean and force deep ocean water, which has been made more buoyant by the downward diffusion of heat, to upwell slowly. Eventually, the upwelled water migrates back to the sinking regions to complete a thermohaline circulation cell (see Gordon et al., 1992). Water masses formed in different regions vary in terms of temperature, salinity, nutrient concentrations, and stored carbon content. The relative contribution from each source region determines the ocean's average temperature, salinity, and other properties, such as carbon storage. In addition, this downward flow of surface water provides a link between the atmosphere and the deep ocean. Thus a better understanding of the global climate system requires a detailed understanding of the thermohaline circulation, its vulnerability to change, and the processes that govern water mass formation rates. Once these factors are understood, they can be represented in global ocean and climate models.

There is evidence that surface salinity fluctuations in the high-latitude North Atlantic Ocean control the thermohaline circulation by altering North Atlantic Deep Water (NADW) formation. One possible mechanism to slow NADW formation is capping of the ocean surface with low-salinity water, such as Arctic Ocean waters (Weyl, 1968). During the last century, there were at least two episodes of low surface salinity water in the northern North Atlantic Ocean [the latter during the late 1960s and 1970s is referred to as the great salinity anomaly (Dickson et al., 1988)] that drastically reduced or stopped convection and NADW production. Changes in the twentieth century are very small compared to suspected changes in NADW formation rates during the swings between glacial and interglacial periods (Boyle, 1990).

The Indian Ocean is a strongly evaporative ocean. Lacking a northern polar region, the tropical heating of the Indian Ocean cannot be vented by flow to the north. Strong evaporation in the Red Sea and Persian Gulf forms warm salty water that, like the Mediterranean outflow, is small in volume but adds significant heat and salt to the deep ocean. The role of the Indian Ocean in larger-scale thermohaline circulation remains unclear and should be studied in the coming years.

The sea ice cover of the Southern Ocean acts to decouple the ocean from the atmosphere, limiting cooling of the ocean by the polar atmosphere. The insulating blanket of sea ice protects the ocean from the cold atmosphere. The extreme seasonality and rapid spring melting of the Southern Ocean sea ice cover suggest that the heat carried into the surface layer by the upwelling of deep water is a key in understanding the Southern Ocean sea ice budget; the buildup of heat within the mixed layer under the winter ice cover induces melting even before solar radiation melts the ice from above. Ocean heat flux also limits sea ice thickness during the winter to less than 1 meter, in contrast to the 3-meter ice of the more stable Arctic Ocean.

Heat and Freshwater Fluxes The ocean interacts with the atmosphere in affecting the heat and freshwater fluxes that control the climate system. Estimates of the fluxes to and from the ocean and atmosphere are difficult to obtain with adequate accuracy, and at the present time there are serious conflicts among values estimated from atmospheric models and data, from oceanic models and data, and from the boundary layers in the two media.

The ocean and atmosphere contribute roughly equally to the transport of heat from lower latitudes toward the polar regions—a transport that is required to maintain the global radiation balance at the top of the atmosphere—although the relative importance of the two transport mechanisms varies with latitude. The massive amounts of water moving in the ocean render it a crucial transporter of moisture, but we have little idea of the sizes of freshwater sources and sinks over the ocean (Baumgartner and Reichel, 1975). Their distribution must have profound effects on the distribution of rainfall over land, an important component of the climate in habitable regions of the world. Low surface salinity caps the ocean, attenuating convection and deep-reaching water mass formation. High surface salinity, combined with surface cooling, allows deep convection and ventilation of the interior of the ocean.

We need to monitor changes in ocean surface salinity globally. Satellite sensors monitor sea surface temperature and its anomalies, but monitoring sea surface salinity and its anomalies on a global scale is beyond present capabilities, as is determination of the temperature below the very surface.

Over the last decade, oceanographers have begun making direct estimates of the north-south ocean heat and freshwater transports using transoceanic hydrographic sections and modern measurements of strong boundary currents and then comparing them with the more uncertain indirect estimates based on atmospheric data. To understand heat and freshwater transport fully, oceanographers must describe the general ocean circulation and its variability. It is discouraging, however, that since 1985, only six transoceanic hydrographic sections, the backbone of the observations needed to determine the north-south fluxes, have been carried out: at 47°, 24°, and 10° north latitude in the Pacific; 32° south latitude in the Indian Ocean; and 11° north latitude in the Atlantic. Overcoming organizational and funding obstacles for these long hydrographic sections takes major effort by individual scientists. However, WOCE has plans to make the necessary measurements for determining the ocean heat transport in each ocean basin at several latitudes (WOCE Scientific Steering Group, 1986). This work is a central focus of WOCE and is clearly needed if understanding of how the present climate system works is to progress.

Provision for continuing such observation beyond the end of WOCE is essential.

Methods

To achieve their scientific objectives, physical oceanographers must use both proven methods and new technologies to make more complete ocean observations.

Volunteer Observing Ships The maritime industry is a resource for ocean research and monitoring that can no longer be viewed as simply an adjunct to the academic research fleet. On the contrary, its integration into a global ocean observing system would provide far greater and more frequent access to the ocean than will ever be possible with research vessels alone.

Volunteer observing ships (VOS) offer opportunities to study and monitor the ocean with a coverage and frequency that are unthinkable by any other means. With the advent of new and more sophisticated remote sensing techniques, such as ocean color scanners, altimeters, and scatterometers, it is plausible that the demand for direct observations will increase, for several reasons. First, the need for calibration measurements will grow. Second, the ocean color scanner will observe numerous signals that require in situ samples for identification, interpretation, and analysis. Third, as coverage of the ocean surface improves, a concomitant need for improved subsurface coverage is inevitable.

Without doubt, a major impediment to the use of VOS is the lack of automated instrumentation for nonscientists on moving commercial ships. Most instruments are designed for trained personnel on research vessels equipped with laboratories. A new approach to the VOS concept is needed. It will require discussions and planning with the maritime industry internationally to develop new modes of cooperation. The community must think of VOS as a potential platform, and ship operators must be persuaded that oceanographic work is to their benefit. At the same time, the development of instrumentation optimized for use on VOS must be encouraged, including the following:

• Modern sensor packages are needed that can be dropped and retrieved repeatedly along a ship's route to measure salinity, oxygen, and fluorescence (primarily from phytoplankton). Data obtained by the sensors could be transferred to a small on-board computer for analysis and transmission to data centers.

• Disposable free-falling sensor packages should be developed that travel to great depths and transmit their data via acoustic signals or return to the surface and broadcast to satellites. Sensor packages could transmit data for any ocean observation that can be measured in situ. Disposable sensor packages can transmit data via a thin wire that detaches when the sensor reaches its maximum depth, but free-falling sensors could travel much deeper and measure more accurately than present devices.

• VOS could tow instruments to measure various ocean characteristics. When ship routes and sampling locations coincide, passing ships could retrieve data from moored instruments (e.g., moored current meters and inverted echo sounders) by acoustic signals. At the least, commercial ships equipped with acoustic Doppler current profilers could measure upper ocean currents and heat fluxes routinely.

Remote Sensing by Satellite   Satellites observe vast areas of the ocean surface daily, obtaining data at a far faster rate than surface vessels and instruments (e.g., Stewart, 1985). Satellites also aid physical oceanography by transmitting data gathered by in situ ocean sensors combined with highly accurate positioning information. The availability of surface and subsurface instruments has greatly increased the volume of data beyond that possible from ship-based observations, and the instruments provide valuable time-series data.

Satellite sensors are used to map natural infrared and microwave radiation emitted from the sea surface. Infrared radiation provides images of sea surface temperature patterns. Radiation in the microwave frequencies is used to map sea ice distribution in areas of the globe that are otherwise poorly accessible. Phytoplankton blooms can be monitored from space with sensors that respond to the visible and near-infrared radiation reflected by plankton chlorophyll, integrating the effects of temperature, nutrients, and the physical structures and processes of the ocean surface layer. Suspended sediment is also measured by visible sensors. Surface winds are measured by satellite scatterometers. The sea surface slope can be measured from space by radar altimeters with accuracies adequate for determining many features of ocean circulation. This is the only practicable method for global-scale continuous observation of circulation. Synthetic aperture radar on satellites can measure sea state, internal waves, and ice conditions with very high spatial resolution.

Although satellite data yield a nearly continuous view of the ocean, it is important to note that complementary in situ observations from ships or instrumented drifters are required to improve the mathematical relationships (algorithms) for calibrating satellite observations to in situ values.

A suite of satellites launched by the United States, the European Space Agency, and Japan and with many joint satellite missions should provide many data required by physical oceanographers during the next decade. Advances in physical oceanography are tied closely with satellite views of the ocean and telemetry of data from Earth's surface. Coordination of satellite projects with ocean science objectives and in situ ocean measurements is required. The satellite time series of sea surface temperature, winds, sea ice, ocean color, and ocean height must be continued without interruption to provide a complete record of their variability.

Tracers Distribution of ocean properties can be used to depict the pattern of ocean currents and the effects of mixing within the ocean because water masses from different source regions vary in their physical and chemical signatures. Traditionally, temperature, salinity, oxygen, and nutrient concentrations have been used to track the movement or spreading of ocean water masses. Other more exotic chemicals present in minute concentrations (tracers) are now commonly used to track water masses. For example, chlorofluorocarbons, carbon-14 and tritium from atmospheric nuclear bomb testing, and other natural and synthetic substances with known rates and times of input to the atmosphere have been used as tracers. Tracers average the spreading action of ocean circulation at a variety of scales and integrate the effects of many processes. The infiltration of naturally occurring and synthetic chemical tracers into the ocean provides insight about the time scales of ocean circulation and mixing. The development of baseline time series of tracer concentrations is important.

Acoustic Techniques Because the ocean is transparent to sound but opaque to light, acoustic techniques provide oceanographers with the opportunity to see the interior of the ocean. In a real sense, the hydrophone array serves as the underwater eyes and ears of the oceanographer. The enormous bandwidth of available underwater acoustic instrumentation (10^{-3} to 10^6 hertz) allows sound to be used as a probe of structures and processes whose scales range from millimeters to ocean basin scales. The ocean is especially transparent to low-frequency sound. Consequently, underwater sound is becoming an important means of studying the three-dimensional structure of the ocean below its surface. Continuous measurements of current velocity from shipboard acoustic sensors give a two-dimensional record of ocean currents to several hundred meters. Acoustic techniques are also used to track sub-surface floats and transmit information between ships and sensors at depth. Yet, sound is an underemployed tool in oceanography. Significant advances should be made during the next decade in physical, biological, and geological oceanography as a result of thoughtful application of acoustic principles and techniques for direct probing and information transfer. Because of its inherent global nature, the acoustic monitoring of ocean climate is a strong candidate for a global ocean observing system.

The air-sea interface provides a variety of challenges and opportunities to the oceanographer using acoustic techniques. The formation and subsequent collapse of bubbles are important sources of sound whose monitoring could provide an estimate of wave breaking intensity from which gas transfer rates could be inferred. Such measurements give important insight into poorly understood sea surface physics. The passive measurement of rainfall-generated sound is a way to measure precipitation in the open ocean. Direct measurement of precipitation is difficult to obtain and generally inaccurate. Better measurements of precipitation over the ocean are important because of its effect on the global heat and water budgets.

Some of the oceanographic applications of underwater sound are simple. Others will require improvements in our understanding of the physics of sound propagation in the sea and improved signal processing techniques and instrumentation.

DIRECTIONS FOR MARINE GEOCHEMISTRY

Summary

Studies of chemicals dissolved in seawater, adsorbed on suspended particles, incorporated in living or nonliving organic material, and buried in seafloor sediments have yielded much information about Earth processes and past conditions. Environmental conditions are imprinted on particles that fall to the seafloor and are buried over time. With adequate understanding about processes that affect chemical concentrations and forms after deposition, sediments recovered by seafloor drilling can illuminate Earth's environmental history for millions of years into the past. In addition, modern ocean processes can be studied by measuring the concentrations of trace elements and compounds in seawater. For example, measurement of trace element distributions is a major tool used by physical oceanographers to study ocean circulation.

The ocean is a chemical reactor, with inputs, internal reactions, and outputs. Inputs are received from continents via rivers and airborne transport. Other chemicals enter the ocean from hydrothermal sources, primarily at mid-ocean ridges. As oceanic crust is subducted beneath continents, elements are expelled from both the crust and its overlying sediment layer. Finally, a minor proportion of input to the ocean comes from cosmic sources. Inputs are the least understood part of the reactor.

Elements are redistributed in the ocean by circulation and mixing, and are transformed through chemical reactions and biological activity. Chemicals finally exit the reactor through incorporation into seafloor sediments. The residence time of chemicals in seawater and in the sediments depends on properties of the chemicals, as well as chemical, physical, and biological conditions. Chemical oceanographers seek to understand the present reactor and then, from examination of changes in the outputs over time, determine variations in the reactor's behavior and the compositions and fluxes of inputs in the past. With this information, the limits of future oceanic changes under given climatic and tectonic scenarios can be estimated.

Future studies in chemical oceanography will be aided by new instruments that are capable of analyzing a wide variety of elements and isotopes contained in small samples. Greater knowledge of processes controlling fluxes, redistribution, and removal, and improvements in our ability to read the sedimentary record are likely in the coming decade.

Introduction

Marine geochemistry integrates several oceanographic disciplines. The aims of marine geochemistry are (1) to understand the inputs of elements from the continents, mantle, and cosmic sources into the ocean over time; (2) to understand the process of material removal from the ocean to the sediments and oceanic crust; (3) to understand the process by which elements and their isotopes are redistributed; (4) to determine the mechanisms of chemical coupling between the ocean and the atmosphere, and interpret the sedimentary record of past oceanic change; and (5) to study marine organic compounds both in their relation to the above factors and to the global carbon cycle, and as detectors of oceanographic properties over time.

Oceanic sediments record environmental events over the past 180 million years of Earth's history. Ocean water in which sediment particles are formed or through which they pass is both an agent of transport and the site where information about environmental conditions is imprinted onto components that eventually become part of marine sediments. Geochemical signatures in the sediments complement the sedimentological and paleontologic record in special ways. Oxygen and carbon isotopes in marine calcareous shells or tests and organic remains provide insight into ocean surface temperatures, ocean circulation, and the extent of ice storage in past glaciations. Trace elements such as cadmium and barium add to our understanding of deep-ocean circulation and upwelling. Isotopes produced by cosmic rays, those originating from uranium decay, and those produced or enhanced by human activities, provide information on chronology, sediment properties, oceanic upwelling, deep-water formation, and transport of elements. The unique properties of the isotope ratios $^{87}Sr/^{86}Sr$, $^{206}Pb/^{207}Pb$, $^{143}Nd/^{144}Nd$, and $^{187}Os/^{186}Os$ permit insights into the geologic questions of plate tectonics because the isotopic composition of seawater reflects the isotopic composition and relative importance of each source, for example, material derived from the mantle and continental rock weathering.

Although present in greatly diluted concentrations, marine organic substances play a key role in the global carbon cycle as modulators and tracers of oceanic processes. For example, organic remains account for approximately 20 percent of all carbon buried in marine sediments and thus are an important sink for atmospheric carbon dioxide. The burial of organic matter and its subsequent oxidation essentially control atmospheric oxygen levels over geologic time. Dissolved organic matter in seawater may contain a mass of carbon comparable to that in terrestrial biomass and could potentially affect atmospheric carbon dioxide concentrations on a time scale of a thousand years—the ocean's turnover time. In addition, organic molecules in particulate and dissolved forms are important vehicles for the transport of reducing power, nutrients, and trace elements throughout the ocean and across the air-sea and sea-sediment interfaces. Organic molecules in the marine environment are couriers of unique information about the sources, pathways, and histories of the associated particles and water.

The Ocean Reactor

The ocean receives dissolved and particulate material from a variety of sources and pathways. Traditionally, river inputs were regarded as the major contributor of dissolved material. Wind transport of dust was recognized as adding to the particle flux borne by streams, and the input of altered volcanic material to deep-sea sediments was clearly identified by the 1960s. Within the past 15 years, deep-sea hydrothermal activity, predominantly at spreading centers, was found to be important as both a source and a sink of elements. In addition, a significant flux of dissolved material may be expelled from sedimentary wedges and underlying oceanic crust as they descend into the mantle through subduction zones. Cosmic dust and cosmic ray-produced nuclides are not a major input to the ocean in terms of volume, but are important as tracers and for understanding episodic influxes of significance to the history of our planet.

Elements are redistributed within the ocean in dissolved form by horizontal and vertical advection, diffusive mixing, and incorporation into particles. Chemical species are removed from the ocean as particles settle to the seafloor; they also react directly with the seafloor as the result of diffusion and circulation through sediments and the oceanic crust. Seafloor sediments and altered rocks undergo secondary reactions over time, and their composition may change substantially. From a chemical point of view then, the oceans are a large chemical reactor with multiple feeds and outputs. Chemical oceanographers want to understand the present reactor and then, from examination of changes in the outputs over time, determine variations in the reactor's behavior and the compositions and fluxes of inputs in the past. With this information, the limits of future oceanic changes under given climatic and tectonic scenarios can be estimated.

Geochemical study of the oceanic water column and the output of elements to the sediments is now relatively mature. The descriptive phase is largely complete, and studies of mechanisms are growing more numerous. For inputs of elements to the ocean, however, quantitative research is difficult and, in the case of inputs from the continents, may not correctly reflect disturbances in terrestrial inputs caused by humankind. These disturbances have occurred on time scales of a few years to several centuries, and the ocean has not yet equilibrated to the altered inputs.

Fluxes

Quantitative measurement of river inputs is difficult because measurements of fluid discharge from rivers are uneven in quality, frequency, and distribution. Because the best data are available from developed regions of the world, they do not necessarily represent areas less impacted by human activities.

Many of these same uncertainties apply to airborne inputs. Transport is strongly seasonal and diffuse, and thus is difficult to measure. Wind erosion rates are sensitive to the nature of the land cover and therefore to changes in land use. Windborne particles efficiently scavenge volatile chemicals released to the atmosphere by volcanism, biomass burning, and industrial activities. Thus the chemical composition of windborne dust is sensitive to pollution. In addition, photochemical reactions occur in the atmosphere, changing the chemistry of the atmosphere and particle-bound chemicals.

Hydrothermal activity in the deep sea is driven by volcanic processes in the oceanic crust, predominantly at spreading centers. Seawater seeps into crustal rocks, entering convection cells in the rocks, and is heated to 300 to 400°C. In the course of this circulation, elements in the seawater react with the hot basaltic rocks; some chemicals are removed (e.g., magnesium, sulfate, and uranium) and others are added (e.g., mobile elements and gases). The fluxes of elements at any one site are difficult to quantify, and it is not feasible to measure the thousands of sites that differ in rock temperature and composition. To account for observed isotope concentrations in seawater (e.g., strontium), a volume of water equal to the world's ocean must circulate through the hydrothermal system at temperatures above 325°C every 10 million years. However, estimates of hydrothermal circulation based on heat lost in the formation of the oceanic crust are about five times lower. If the higher estimate for hydrothermal circulation is correct, this process is as important as rivers for the input of many elements to the ocean. Hydrothermal processes would stabilize seawater composition and thus act as a geochemical flywheel, potentially damping large-scale changes induced by long-term climatic and tectonic changes. Yet, if the lower estimate of hydrothermal circulation is correct, hydrothermal activity is a minor factor in the cycling of ocean elements and is important for only a few. The major inconsistency between the fluxes based on isotopic and thermal constraints, apparent since the first hot springs were found in the deep sea 15 years ago, remains to be resolved.

There is evidence that hydrothermal circulation at relatively low temperatures (a few tens of degrees) away from spreading centers may also be important for fluxes of elements. However, it is not yet possible to calculate even the vaguest estimate of the chemical fluxes involved. Recent seafloor exploration and ocean drilling show that the ocean crust and its overlying sediment expel water in the subduction process. Given the great compositional and tectonic diversity of active subduction zones, it will be impossible to estimate the geochemical significance of these processes to the oceanic element cycles until much more exploratory work has been done.

Redistribution

All the inputs will react with seawater. Reaction processes in the zone of mixing between river water and the upper ocean are quite well described qualitatively. These processes include desorption of elements from suspended particles, coagulation and precipitation of colloidal material, scavenging by organisms, and vertical transport. Processes associated with the formation of particle plumes above hydrothermal vents have been studied extensively. These particles oxidize rapidly and appear to scavenge both a large proportion of the hydrothermically transported trace metals and a significant component of elements from ambient seawater.

As mixing progresses, ocean circulation is increasingly dominant in dispersing inputs until they cannot be traced directly back to their source. Large rivers can be considered as point sources of material to the ocean superimposed on a diffuse background input from smaller streams. Penetration of the river signal into the deep sea follows complex pathways that are regionally diverse, depending on the current regime and the configuration of the coastline and the continental shelf. Because even unpolluted rivers usually carry elevated nutrient loads relative to coastal seawater, their discharge induces large phytoplankton blooms. The phytoplankton settle toward the seafloor, carrying nutrients and scavenging dissolved substances. In confined systems with high nutrient inputs, settling organic material can fuel bacterial activity and lead to oxygen depletion of bottom waters. The complex coupling of inorganic and biological processes, postdepositional reactions in the sediments, and strong seasonality of inputs produce a system whose chemical transport is difficult to quantify. Estuarine processes make it difficult to estimate river inputs of the more reactive elements to the open ocean. Because of this complexity, chemical oceanography in the coastal ocean has been relatively neglected even though that is the site for some of the most intense biogeochemical interactions in the entire ocean.

High-temperature hydrothermal fluids create great buoyant clouds of fine-grained sulfides and oxides upon their turbulent injection into the surrounding water column. These clouds rise above the vents until their density equals that of the surrounding seawater. On ridges with high rift walls (e.g., the Mid-Atlantic Ridge), the plumes often stay within the confines of the bounding rift walls. The precipitates and dissolved material they scavenge then accumulate on the walls and floor of the rift valley. On fast-spreading ridges with low walls (e.g., the East Pacific Rise), the plumes rise above the walls and are dispersed in the middepth circulation. Fine-grained vent-derived particles can be transported for thousands of kilometers, slowly settling out to form a compositionally distinctive shadow in the sediments. Unreactive species, such as helium, delineate plume flow.

If redistribution of elements occurred only by the physical dispersal of dissolved material and suspended particles near their sources, geochemical patterns would be related solely to input functions. However, the chemistry of the ocean reactor is determined primarily by organisms and chemical kinetics rather than by thermodynamics. Essential nutrient cycles are controlled largely by the metabolic processes of living organisms. For nonnutrient elements, scavenging by nonliving organic materials is much more important. Settling particles are reactive enough to adsorb and transport a large variety of elements to the bottom. Because bacteria continue to degrade this material, the concentrations of many elements increase with depth and along current flow lines.

Surface productivity shows strong regional variability. The vertical particle flux and the intensity of scavenging and release also vary by region. For geochemical purposes, satellite pictures of ocean productivity must be projected into the vertical dimension to appreciate the fact that strong lateral variations in reactivity occur (reactivity is the intensity of the chemical reactions that are driven by biological activity in the ocean water column). Vast areas, such as the subtropical gyres, are quite unreactive; these are surrounded by a coastal rim of high reactivity. Other zones of high reactivity correspond to areas of physical upwelling.

The combination of lateral and vertical transport and continuous reaction of particles suggests that the water column distribution of elements at a particular location may have little influence on local processes but, instead, reflects the integration of processes over various time scales and distances. The best example of this effect is the deep silica maximum found throughout the Indian and Pacific oceans. It results from strong upwelling and associated high productivity of siliceous plankton in relatively small areas at these oceans' northern boundaries, and subsequent vertical transport and dissolution of their siliceous tests at depth.

The flux of particles to the seafloor is apparently the most important mechanism by which elements are delivered to the sedimentary reservoir. However, many complex processes, primarily biological, can occur after initial deposition. Particles falling from above are the only source of nutrition for many bottom-dwelling organisms. Thus this material is ingested and excreted repeatedly by a variety of species and is also subject to continuous bacterial degradation, reducing its carbon content and destroying the functional groups responsible for scavenging reactive elements from the water column.

Sinks

The two most important properties of the ocean system that control the uptake of chemical species by the sediments and oceanic crust are its oxidation state and its temperature.

Chemical reactions in sediments involving oxidation and reduction depend on the amount of oxygen present, which in turn depends on metabolic activity and diffusion from the overlying water column through the pore fluids. Where metabolic processes exceed the oxygen supply through diffusion, the sediments become anaerobic. This condition creates diffusion gradients in the oxic-anoxic transition zone from above and below, affecting a range of chemical reactions. The transition zone rises through the sediment column as sediment accumulation progresses. Understanding and quantifying these processes are crucial for studies of global change because of their implications for interpreting the sedimentary record.

The character of the crustal sink for oceanic dissolved material changes with the temperature of the water-rock interactions. At high temperatures, the reaction environment is anoxic. Sulfide-forming elements are precipitated; elements that form insoluble oxides in the reduced state, such as uranium and chromium, are also precipitated. Soluble materials, such as boron and alkali compounds, are completely removed from the rocks. Magnesium is removed but calcium is released. At the very high temperatures (>400°C) associated with recent eruptions, phase separation can occur, producing a dilute aqueous phase and a residual brine. Subsequent mixing of these components appears to be responsible for the large variations in salinity observed in hydrothermal flu-

ids. Details of this process are not understood. A major unknown is the mechanism responsible for maintaining the salinity
and temperature of the vent fluids at a given site at stable levels
over periods of years.

Low-temperature weathering of the oceanic crust appears to
constitute a major sink for alkali compounds and is accompanied
by extensive hydration of the rocks. When the crust is subducted,
this chemically bound water, along with the elements it can transport,
most likely is released and migrates. The water may accelerate
melting and participate in the eruption process in volcanic arcs.

Marine Organic Substances

The biochemicals that fuel marine organisms are photosyn-
thesized and then respired in the upper ocean on time scales of
hours to days. Only about 20 percent of the photosynthetic product escapes from the sunlit surface ocean as sinking particles, and
less than 0.5 percent is ultimately preserved in marine sediments.
Living organisms comprise only about 1 percent of the organic
matter in the ocean. The remaining organic matter is primarily a
dilute solution (about 1 part per million) of "dissolved" macro-
molecules (i.e., material that passes through filters with a pore
size of 0.5 micrometer). The turnover rate of this dissolved pool
is now under discussion; the traditional view is that the pool
turns over at a rate of thousands of years. The alternative view is
that, because the pool of dissolved organic material contains excess carbon-14 relative to what is expected in thousand-year-old
organic material, it must turn over more rapidly. Because of the
challenges of isolating or directly characterizing this extremely
dilute component of seawater, only about 20 percent of the organic molecules have been described.

A little over a decade ago, a novel suite of organic lipids was
first reported in sediments from the Atlantic Ocean and the Black
Sea. The component molecules have a linear sequence of 37 to 39
carbon atoms containing one to four double bonds, with an oxygen atom doubly bonded to the second or third carbon in the
chain. These long-chain alkenones were found to be produced by
the marine coccolithophorid algae *Emiliania huxleyi* and related
species that are widely distributed in tropical and subtropical oceans.
The same molecules were also discovered in sediments dating
back to the Miocene (about 20 million years ago).

It was later demonstrated in the laboratory that the average
number of double bonds (extent of unsaturation) in these alkenones

of *E. huxleyi* and related algal species increased at lower culturing temperatures, apparently a chemical response to maintain membrane fluidity. Brassell et al. (1986) discovered that in natural deposits the average number of carbon double bonds in alkenones from Quaternary sediments of the northeast tropical Atlantic showed a strong inverse correlation, over the past 120,000 years, with the temperature of the near-surface ocean water as inferred from the stable oxygen isotopic composition of coexisting calcium carbonate shells of the planktonic foraminifera *Globerigerinoides sacculifer.* This observation suggests that the relative changes in water temperature at the ocean surface during at least the past 100,000 years could be inferred from the stratigraphic record of alkenone composition in the underlying sediments. Prahl and Wakeham (1987) calibrated the alkenone paleothermometer for *E. huxleyi* within ±0.5°C and they provided evidence against the idea of diagenetic alteration of the molecular temperature record in the marine water column.

The alkenone paleothermometer has potential applications beyond simple confirmation of stable oxygen isotopic records of sea surface temperatures. For example, alkenone measurements can be applied readily to bulk sediment samples. The alkenone paleothermometer thus ranks as one of the major contributions in the past decade of marine organic chemistry research to understanding paleoceanography and paleoclimatology.

The application of the alkenone paleothermometer was made possible by the recent development of gas chromatographic ratio mass spectrometers, instruments that measure the stable carbon isotopic compositions of individual organic molecule types separated during the rapid (approximately one hour) gas chromatographic analysis of complex organic mixtures.

Future Directions

Analytical Methods

The chemical processes occurring within the ocean reactor are kinetically controlled except in high-temperature regimes, where thermodynamic equilibria may be inferred. These kinetic processes are driven largely by organisms; they involve chemical reactions that are difficult to reproduce in laboratory experiments because of their complexity and variety. Thus, the general strategy in studying the marine geochemistry of the present ocean and its variations in the geologic past is to pose questions in which causes are inferred from measured distributions, that is, as an inverse problem. Variations in measured concentrations reflect changes in the relative importance of these causes. This inversion is constrained by different chemical properties of the elements of the periodic table, their valence states, their isotopes, and the compounds they form. It follows that this strategy requires the measurement of a vast array of chemical properties. Although technology development is required for measuring some of these properties, each of the elements and forms is not equally important diagnostically. Extraction of all the information useful in constraining the inversion is a prodigious analytical task.

Until recently, each available instrument could analyze only a few elements at the sensitivity, precision, and accuracy levels necessary for their concentration in natural samples. Thus no laboratory could perform more than a small fraction of possible measurements. Over the past 10 years, the sensitivity of instruments has increased vastly. Accurate and precise multielement analyses on single samples are now feasible, and multiple collector thermal ionization mass spectrometers have increased the sensitivity for a wide range of elements and isotopes. High-energy accelerator mass spectrometry allows the measurement of the cosmogenic radioisotopes in the study of a wide range of geochronological questions; it also allows exploitation of the unique properties of these isotopes in a variety of tracer studies. Plasma source mass spectrometry makes it possible to perform accurate multielement analyses on extremely small amounts of material. This technology eases measurement of refractory elements that, because of their low volatility or high ionization potential, are difficult to measure with conventional techniques.

Mass spectrometry for the measurement of $\delta^{18}O$, $\delta^{13}C$, and other light isotope systems are well established, as are methods for determining radiogenic isotopes. New developments in high-resolution thermal ionization mass spectrometry for the measurement of ^{230}Th, ^{232}Th, ^{234}U, and ^{238}U and the negative thermal ionization mass spectrometric measurement of $^{187}Os/^{186}Os$ have improved our capacity to use these isotopes in marine geochemical studies.

Improving Our Knowledge of Fluxes

Much more research needs to be done to quantify continental inputs to the oceanic reactor. An important adjunct to studies of fluxes will be long time-series measurements that will allow an estimation of variability of fluxes. Weathering processes are traditionally regarded separately from oceanography, but oceanic techniques are essential to trace the pathways by which material enters estuarine mixing zones through rivers, progresses across the shelf, and finally moves into the deep sea. For some elements (e.g., iron), delivery to the world ocean by this route is negligible because of their insolubility in seawater. Soluble elements pass through the coastal ocean with little loss. Research on this topic has great societal importance because chemical fluxes substantially alter global climate, coastal pollution, and possibly harmful algal blooms and fisheries production.

The strength of airborne transport of particles is known to vary widely with climatic conditions. Airborne transport is the one direct pathway between the continents and the surface waters overlying the deep ocean, and its importance as a supplier of micronutrients (e.g., iron and selenium) needs to be more firmly established. It is possible that the productivity of some areas of the ocean is controlled partially by the amount of airborne trace metals. This area of research is an active one.

Estimates of the hydrothermal flux range over a factor of five. The mechanisms responsible for the wide salinity variations and the temporal stability of the values at a given site are not understood. Both of these problems can be resolved only by systematic investigation of vent fluids from different sites and by development of additional tracers of both the subsurface reaction processes and the characteristic hydrothermal inputs to the global reactor.

Fluid inputs from the mid-ocean ridge flanks and from subduction zones are perhaps best studied by drilling and pore water sampling because the flow across the seafloor-water interface may be too diffuse for discrete sampling in the water column. Much improved down-hole sampling and measurement capabilities are required. Systematic sampling of representative ridge flanks and subduction zone complexes is needed, using the complete range of modern geophysical tools.

Understanding Redistribution and Removal in the Ocean

The general circulation of the global ocean has been relatively well described. From the geochemical point of view, much remains to be learned about the relative importance of the removal of elements at ocean boundaries versus in situ removal by settling particles. The spatial and temporal variability of processes controlling the vertical flux of elements needs to be characterized. The same is true for the evolution of the chemical properties of sinking particles traveling through the complex food web in the water column and on the seafloor. In the upper waters and thermocline, large horizontal variations in primary productivity and higher levels of food webs have been observed, but little is known about the lateral variability of deep-ocean ecosystems. The effects of ecosystems on the chemistry of particles passing through them are largely unknown. It is important to know whether there is a simple proportionality between surface productivity and the chemistry of the underlying water column or whether the particular faunal assemblages in the water column exercise a major role in controlling element concentrations.

The mechanisms of sediment interactions and diagenesis are well studied for the major constituents, nutrients, and oxygen; much more work is needed on the behavior of minor and trace constituents. Intensive research on the behavior of trace elements in the most reactive upper few meters of the ocean is also necessary. Descriptions of composition changes need to be developed for the various sediment types and environments through all phases from initial burial to subduction.

Reading the Record

Complete decipherment of the proxy record contained in the concentration distributions of trace elements and isotopes in sediments requires an understanding of the pathways of input and the mechanisms of redistribution, removal, and transformation of the elements studied. A given tracer may record a single aspect or some combination of these factors. Ideally, multiple tracers should be used as a check on internal consistency. Because only a small number of tracers are now available, much development work is required.

It is known that the continental inputs of a number of potential tracers are changed markedly by human influence, which makes estimates of their response to environmental changes difficult. River inputs are often strongly mediated by coastal processes and hence are sensitive to sea-level variations that change the size and character of the coastal ocean. Comparative studies of shelf-dominated systems (e.g., the Yangtze and Yukon rivers) with systems in which rivers discharge directly over deep waters (e.g., the Columbia and Congo rivers) may be informative.

The relative importance of physical and chemical redistribution processes and the mechanisms of sediment uptake must be established for each tracer; this will be difficult experimental work. Postdepositional redistribution processes must be characterized for the range of sedimentary environments. The establishment of reliable proxy records merits high priority because it is the most direct constraint on the modeled mechanisms of previous global changes in environmental processes.

DIRECTIONS FOR MARINE GEOLOGY AND GEOPHYSICS

Summary

The plate tectonic paradigm, first quantitatively described more than 25 years ago, provides an integrated physical and chemical framework for understanding the geological evolution of Earth. Marine geologists and geophysicists played a critical role in the development of this paradigm. By linking marine magnetic anomalies to geomagnetic reversals of Earth's magnetic field, marine geophysicists were able to confirm seafloor spreading and provide quantitative estimates of seafloor spreading rates. Through holes drilled by the Deep Sea Drilling Project, marine geologists were able to extend the geomagnetic reversal time scale back nearly 200 million years, providing a framework within which to reconstruct the past positions of the continents and the opening and closing of ocean basins.

Throughout most of the 1970s, the major emphasis in marine geology and geophysics was on large-scale kinematic descriptions of relative plate motions and their consequences for the geological evolution of ocean basins. However, by the 1980s, the focus of the field had shifted toward more process-oriented studies centered around understanding how oceanic crust and lithosphere are created, how these processes are related to the underlying mantle, and the consequences of seafloor spreading on ancient ocean circulation and climate. Four themes currently dominate research in marine geology and geophysics: (1) the formation of oceanic crust and lithosphere along mid-ocean ridges, and the associated volcanic, hydrothermal, and biological processes; (2) off-ridge processes and their relation to mantle convection; (3) the structure and tectonics of active and passive continental margins; and (4) the record of past climate change and ocean circulation preserved in marine sediments. In addition to these four themes, there is increasing interest in the study of coastal processes including sediment erosion and transport, and estuarine and delta sedimentation. Coastal processes are discussed elsewhere in this report.

In the 1990s, marine geologists and geophysicists will begin a major, decade-long study of the global mid-ocean ridge system through the Ridge Inter-Disciplinary Global Experiment (RIDGE) program. The long-term goal of this program is to obtain a sufficiently detailed spatial and temporal definition of the global mid-ocean ridge system to construct quantifiable, testable models of how the system works, including the complex interactions among the magmatic, tectonic, hydrothermal, and biological processes associated with crust formation. Among the goals that are achievable in the next decade are a *global* characterization of the structure and energy fluxes along the entire 50,000-kilometer-long mid-ocean ridge system and the establishment of a permanent seafloor observatory on an active ridge segment to investigate the scales of variability in tectonic, magmatic, hydrothermal, and biological processes associated with the formation of new oceanic crust.

The ancient ocean crust contains unique information about mantle convection and composition. An improved understanding of the chemical and isotopic record of mantle convection and the variation of melt production through time is likely in the next decade. Mapping crustal composition on an ocean basin scale will require hundreds of shallow holes to be drilled into ocean crust. Directly sampling the suboceanic mantle will require the development of new drilling technology beyond that presently available to the ocean drilling community. New seismic tomography techniques for imaging the Earth's mantle will allow marine geologists to begin to relate mantle convection processes to melt production rates, lithospheric stress and intraplate deformation, and the variation in chemical and isotopic composition of the crust. Vastly improved seismic images of the suboceanic mantle are possible if an array of seafloor seismic stations is established in the 1990s to augment the global digital seismic network.

Continental margins are the locus of lithospheric deformation, sediment accumulation, and substantial and chemically distinctive magmatism. Subduction and rifting processing at margins determine the size, shape, and distribution of continents and result in complex and dynamic interactions among oceanic crust, continental crust, and mantle systems. A basic description of the nature and evolution of many margins is available today, but understanding of the dynamics of margin development is still very limited. The development of new technology for probing the deeper structure of continental margins, and new conceptual advances in areas such as sequence stratigraphy and fault dynamics, provide an opportunity for the development of fundamental new insight into margin structure and evolution. What is needed is a coordinated, interdisciplinary research effort involving both land-based and sea-based research programs over the next decade.

Marine sediments provide an important record of geological processes including past global climates. For example, analysis of marine sediment cores has provided critical information on the importance of Earth's orbit in short-term climate change. Marine sediments also provide a record of global sea-level changes, sea surface and bottom water temperature variations, changes in ocean current patterns, the volume of water locked in polar ice caps, and the effects of a changing physical and chemical environment on the evolution of marine life. Through drilling and coring, especially in high-latitude regions, paleoceanographers are poised to make major advances in our understanding of the natural variability in global climate systems in the coming decade.

Introduction

The plate tectonic paradigm forms an integrated and linked physical and chemical framework for the flow of energy and mass through Earth (Figure 3-1). Radioactive decay of material within Earth's interior produces heat and creates a convective system that transports heat and material from deep within Earth to shallow levels. Upper mantle rocks partially melt, producing basaltic magma. Much of this melt is preferentially focused along the world-encircling mid-oceanic ridge, where oceanic crust is created. In time, the oceanic lithosphere (the oceanic crust and upper mantle) will be recycled into the mantle at convergent margins. The oceanic crust and some sediment are carried back into the mantle, and the crustal components dehydrate, pumping water and gases into the overlying mantle, causing partial melting and fractionation, and creating silica-rich rocks, ore bodies, and explosive volcanism along the overlying volcanic arc.

As oceanic crust ages and moves away from the ridge axis, it modifies Earth's environment. The chemistry of seawater is altered as the oceanic crust cools and exchanges elements with the seawater that circulates through it (see "Directions for Marine Geochemistry"). At convergent margins, some sediment is scraped off the subducting crustal plates, injecting fluids rich in dissolved constituents into the overlying ocean waters. Moreover, the aging oceanic lithosphere serves as a repository for sediments that

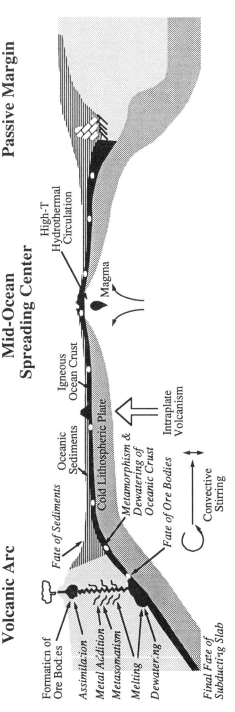

FIGURE 3-1 Cartoon of the solid Earth geochemical cycle showing some of the fluxes, processes that control the fluxes, and the sedimentary reservoirs that provide a record of these processes. Lithosphere is created at mid-ocean spreading centers from upwelling mantle material and is recycled into the mantle at subduction zones. This solid Earth geochemical cycle controls the flux of heat and mass from the Earth's mantle to the hydrosphere, biosphere, and atmosphere.

represent a coherent high-resolution continuous record of environmental changes on time scales from years to millions of years. Given the mobility of crustal plates, the geometry and location of ocean basins and continents have changed through time. This changing plate mosaic has had a profound effect on global sea level and on ocean and atmospheric circulation.

In essence, plate tectonics is the surface expression of a solid Earth geochemical cycle. An understanding of how this cycle has operated in time and space is a fundamental starting point for Earth systems research, and it unifies the field of marine geology and geophysics. The four principal elements of this research are (1) the oceanic ridge and lithosphere, (2) off-ridge processes, (3) ocean margins, and (4) ocean basin sediments.

Oceanic Ridge and Lithosphere

The global mid-ocean ridge is perhaps the most striking feature on the solid surface of our planet. Sections of the ridge extend along the floor of the world's ocean to a length in excess of 50,000 kilometers. The mid-ocean ridge dominates Earth's volcanic flux and creates an average of 20 cubic kilometers of new oceanic crust every year. Two-thirds of the annual heat loss from Earth's interior occurs through the generation and cooling of the oceanic lithosphere, partially by the circulation of seawater through fractures in the hot oceanic crust. This hydrothermal circulation facilitates a major chemical exchange between seawater and oceanic crustal rocks that acts as an important regulator of the chemistry of the ocean and of the volatile content of Earth's interior. The most stunning manifestations of this circulation are the high-temperature hydrothermal vents along the ridge axis.

Many discoveries of ridge phenomena have been made over the past decade, and a number of sophisticated technological tools have been developed for detailed investigation of the seafloor and the subsurface crust. High-temperature hydrothermal vents, for example, were discussed only as theoretical possibilities before their discovery in the Pacific in the late 1970s. High-resolution swath mapping and side-scan sonar imaging systems have only recently begun to provide information on the detailed morphology and structure of ridge systems. Multichannel seismic imaging techniques have advanced and thus have enabled marine geologists to begin imaging the magma chambers that lie below the ridge axis (Detrick et al., 1987). Much of the promise of this new technology remains to be realized. Detailed sampling and mapping of the mid-ocean ridge, for example, have been confined to only a small fraction of the total ridge length. The diversity of volcanic and tectonic processes manifested along the ridge axis, as a consequence, has not yet been fully defined. More fundamentally, the complex and linked processes of magmatism, hydrothermal circulation, development of vent ecosystems, and lithospheric evolution are only dimly understood. The dynamics of these processes have not yet been elucidated because of the lack of in situ observations of sufficient duration and diversity to determine the important interactions and time scales.

For a better understanding of mid-ocean ridge processes and their impacts on the chemical, physical, and biological evolution of the oceanic mantle, crust, and hydrosphere, specific aspects of the ridge system will require focused research efforts. Some are discussed below.

Mantle Flow, Melt Generation, and Magma Transport
Beneath Mid-Ocean Ridges

Plate spreading and the generation of new oceanic crust and lithosphere along oceanic spreading centers involve a variety of complex and interrelated geodynamic processes: upwelling and horizontal divergence of the solid mantle beneath spreading centers, pressure-release melting of this upwelling mantle and segregation of the partial melt from the deforming solid matrix, the emplacement and solidification of melt at shallow depths to create the oceanic crust, and the cooling of the crust and mantle to form the oceanic lithosphere. These processes are still among the more poorly understood aspects of the seafloor spreading process. Two of the most important questions are (1) the pattern of mantle flow beneath mid-ocean ridges, and (2) the geometry of the melting region in the mantle and how melt migrates to the ridge axis. Simple plate-driven flow, due to viscous coupling of the asthenosphere to the separating lithospheric plates, predicts a simple two-dimensional upwelling pattern more than several hundred kilometers in width. Pressure-release melting of this upwelling mantle is thus expected to occur over a very broad region beneath mid-ocean ridges. One of the first-order paradoxes in our present understanding of mid-ocean ridge geodynamics is how partial melt formed over such broad regions beneath ridges migrates to the extremely narrow (1- to 5-kilometer-wide) zone of eruption observed at mid-ocean ridges. We also have only a very crude un-

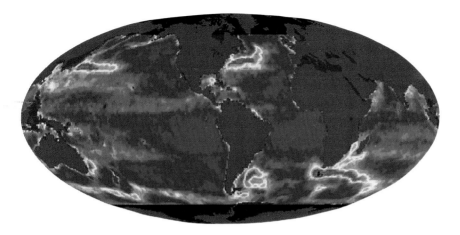

PLATE 1 Average global sea level variability for 1987 and 1988. Data were obtained by the Geosat satellite altimeter. Figure provided by C. J. Koblinsky, NASA.

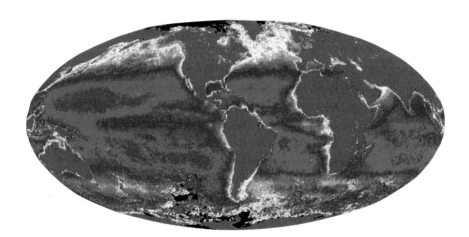

PLATE 2 Multiyear composite of global ocean pigment concentration (November 1978 to June 1986). Data were acquired by the Coastal Zone Color Scanner on the Nimbus-7 satellite. Purple and blue areas contained low concentrations of pigment in surface water and yellow and red areas indicate high concentrations of pigment. Figure provided by Gene Feldman, NASA.

PLATE 6 Integration of GLORIA data with bathymetric data has proven extremely effective in visualizing seafloor geology. This photo shows a perspective view of a portion of the Florida Escarpment with the small ravines cut into it and the meandering channel running across the abyssal plain floor at the base of the escarpment. Photo courtesy of Dr. David Twichell, U.S. Geological Survey, Woods Hole, Massachusetts.

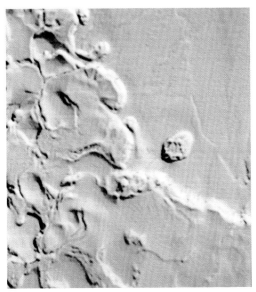

PLATE 7 Shaded relief image of Seabeam bathymetry along the Texas-Louisiana continental slope. Data collected by the NOAA National Ocean Survey. This image shows part of the Sigsbee Escarpment, several collapse basins north of the escarpment, and one salt diaper south of the escarpment. Area shown is approximately 50 by 60 nautical miles. Photo courtesy of Dr. David Twichell, U.S. Geological Survey, Woods Hole, Massachusetts.

derstanding of how this two-dimensional plate-driven flow develops into a more three-dimensional upwelling pattern along some ridges and how this flow affects, or is affected by, the observed segmentation of oceanic spreading centers.

By deploying an array of ocean bottom seismometers across a section of the mid-ocean ridge and recording a sufficient number of seismic events at different ranges and angles, it should be possible to improve the resolution of the seismic structure of the shallow mantle beneath a ridge crest. A major goal for the next decade is to carry out one or more of these studies on the mid-ocean ridge.

Processes That Transform Magma into Oceanic Crust

The transformation of magma into oceanic crust at spreading centers has fundamental implications for the mechanisms of heat and material transport from deep within Earth to the lithosphere, hydrosphere, and biosphere. The important processes that transform mantle melt into oceanic crust and the role of crustal chambers are poorly understood. The global distribution and physical properties of magma chambers at oceanic ridges and their temporal and spatial variability should be determined. Internal dynamics of magma chambers are important factors that must be understood, along with their effects on the structure and composition of the crust, the transfer of heat from the magma chamber, and the physical and chemical processes occurring at the interface between the magma chamber and the overlying region of seawater circulation.

Processes That Control the Segmentation and Episodicity of Lithospheric Accretion

The use of new technology, such as satellites, swath mapping, and side-scan sonar, has revealed that the global rift system is segmented and that the pattern of segmentation varies temporally and spatially. It is essential to understand the physical processes controlling segmentation and its temporal and spatial variation as well as the processes causing episodic production along individual segments and their boundary zones. Melt migration and eruption, faulting, fissuring, and stretching must also be better understood so that the individual processes and their possible interactions can be studied and interpreted.

Physical, Chemical, and Biological Processes
Involved in Interactions Between Circulating Seawater
and the Lithosphere

Hydrothermal plumes that issue from seafloor vents link the oceanic lithosphere, hydrosphere, and biosphere through complex physical, chemical, and biological interactions (Rona et al., 1986). A detailed understanding of the individual processes that constitute a hydrothermal system will provide insight into many problems in biological, chemical, geological, and physical oceanography. Although present research on seafloor hydrothermal circulation has begun to address a few of these problems, new approaches and a more focused effort will be required to achieve an interdisciplinary view.

Distribution and Intensity of Mid-Ocean
Hydrothermal Venting

The character of hydrothermal plumes is determined by both crustal processes and the oceanic environment. Changes in the plume can reflect events with diverse spatial and temporal scales, such as magma chamber evolution, changes in the subsurface hydrothermal plumbing, and shifting bottom currents. To understand these complex interactions, we must study hydrothermal plumes over a wide range of scales in time and space: from the scale of the individual vent plume fluctuating over a period of seconds up to the 1,000-kilometer scale of the large ocean-basin plumes estimated to contain the integrated output from 100 years of hydrothermal venting. An important new research direction is to move from the realm of general observation to the quantification of rates and processes in hydrothermal plumes.

In summary, an improved understanding of the mid-ocean ridge system will require focused efforts. Present technologies are relatively well developed for establishing the occurrence of spatial variations within the ridge system, but obtaining observations of temporal change will be challenging. Global-scale reconnaissance surveys can help in selecting sites for more focused regional studies in which coordinated experiments would involve a range of long-term measurements. A common requirement of many of the recommended studies is accurate age information on time scales between a decade and a million years. Innovative approaches to dating hydrothermal fluids, rocks, and biological materials will be necessary to meet this requirement.

Off-Ridge Processes

Off-ridge processes can be studied to determine how Earth functioned in the past and whether there are additional temporal variability and forcing functions that will not be discovered by studying the geological present. The seafloor contains a record of the creation of the oceanic lithosphere. In addition, important questions concern changes in the oceanic crust as it ages. The older oceanic crust also contains information concerning mantle convection and composition. Several important investigative themes can be identified.

Chemical and Isotopic Record of Mantle Convection

Because the mantle is overlain by the crust, it is not possible at present to sample the suboceanic mantle directly, except in tectonically anomalous areas (e.g., oceanic fracture zones). The basalts that are derived from the mantle, however, are indirect mantle samples that have been modified by partial melting and partial crystallization. Because the oceanic crust is thin and its composition is similar to the magma, and because the spreading center provides a relatively permeable and free pathway to the surface, ocean ridges are the sites where the magma is least modified. Thus ocean ridge basalts typically provide the least adulterated record of mantle composition and temperature. Mapping crustal composition can provide quantitative information about the size, distribution, and composition of mantle reservoirs and the efficiency of convective stirring. This information is a record and an opportunity to map indirectly the composition and temperature of the mantle.

Variation of Melt Production (Convective Vigor) Through Time

There is strong evidence that plate separation rates and basaltic magma production rates along ridges and within plates are not constant. For example, a 50 to 75 percent increase in the rate of formation of oceanic crust and a doubling in the production rate of basaltic magmas between 120 and 80 million years ago (Figure 3-2; Larson, 1991) has been documented. The changes may be due to a large mantle-derived super plume that may have lifted off the core-mantle boundary and have been responsible for increased seafloor spreading and large-scale oceanic plateau production (e.g., Ontong Java and Kerguelen plateaus). It has been suggested that the super

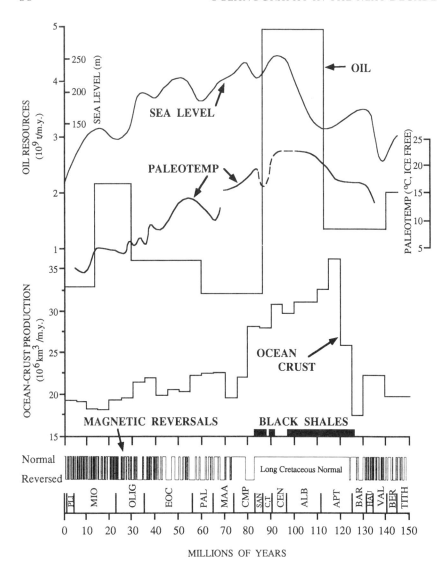

FIGURE 3-2 Combined plot of magmatic reversal stratigraphy; world crustal production; high-latitude, sea surface paleotemperatures; long-term eustatic sea level; times of black shale deposition; and world oil resources plotted on the geologic time scale. Note that increased volcanic activity in the Cretaceous is associated with eustatic sea levels, high sea surface temperatures, and black shale production. Thus, this may be a link between mantle processes (volcanism) and global climate. (Compiled from a variety of sources; Larson, 1991.)

plume had major geological consequences, including considerable increases in eustatic sea level, paleotemperature, oil generation, black shale deposition, and species diversification of phytoplankton and zooplankton. Such an event has profound implications for our understanding of mantle dynamics, oceanic plateau formation, and global environmental change.

Structure and Composition of Oceanic Crust

An understanding of crust formation cannot be achieved in the absence of better knowledge of the composition of the total ocean crust. This information would allow solution of a host of long-standing controversies, including the relationship between crustal structure and spreading rate, the origin of the seismically defined stratigraphy of the oceanic crust, the total magnetization of the crust and how it is distributed with depth, and the depth and nature of hydrothermal interaction in the crust.

Knowledge of Stresses Acting on Oceanic Lithosphere and Intraplate Deformation

The observational basis for plate motions is well established, but the relative importance of the forces (ridge-push, trench-pull, and plate-drag) that act on the plates and cause them to move is unresolved. In addition, the stresses that act upon the oceanic lithosphere at or near plate boundaries are poorly understood. Determining the stresses required to create these structures is key to understanding the tectonics of these environments (Zoback et al., 1985).

Ocean Margins

Continental margins are a principal site of lithospheric deformation, sediment accumulation, and mass flux on Earth and the site of substantial and chemically distinctive magmatism. Understanding their nature and origin will provide knowledge of the history of the ocean basins, and because the margins are progressively incorporated into the continental mass by plate interactions, the knowledge is also essential to our understanding of the mechanisms of continental evolution. In the next several years, the opportunity exists for researchers to develop a fundamentally new understanding of margins, a leap that may parallel that brought about by the plate tectonic revolution some 25 years ago.

A basic description of the nature and evolution of many continental margins is available today. However, understanding of the dynamics of margin development has not grown at the same rate. Knowledge of continental margin structure and motion has progressed, but understanding of the margins is limited by the need to understand the basic physical processes that accompany margin formation. Continental margins research must embrace a unified approach that emphasizes the important role of process-oriented interdisciplinary programs. Further, margins cross the shoreline, and such efforts will further enhance the developing synergism with Earth scientists involved in land and ocean research.

Fault Dynamics and Lithospheric Deformation

Any approach to the problems associated with lithospheric deformation demands the use of many tools. A wide-ranging field-mapping program is needed specifically to establish the link between surface deformation and deformation in the lower crust. For example, the configuration of large faults at depth in the crust and the structural fabric associated with distributed deformation in the upper mantle must be established by large-scale seismic imaging and tomographic techniques. Laboratory studies are also essential to establishing the constitutive laws for frictional behavior of both large deformation zones and discrete fault zones and defining their variability with strain, strain rate, and fluid content. Models of fluid flow in deforming porous media and methods to relate the models to observable quantities for field observation are required.

Mantle Dynamics and Extension

Understanding the interaction between mantle processes and lithospheric extension will require a focused multidisciplinary effort. Wide-ranging field-mapping programs must be designed to establish specific links between surface deformation and deformation in the lower crust. Large geophysical experiments will be needed to define the modern structure of rifts and margins and the thermal and dynamic state of mantle beneath young, presently active rifts. This work must be combined with geological mapping and thorough geochemical studies of magmatic systems to determine the nature of mantle sources, the history of melting, and fractionation. The development of shear-wave techniques and electromagnetic methods to increase their sensitivity to the detection of fluids will be particularly relevant.

Recycling

Mass flux studies needed to understand convergent margins as dynamic systems will involve integrated field, analytical, experimental, and theoretical studies that draw from geochemistry, petrology, and geophysics. Such projects require marine as well as land-based studies and careful integration of results from the two. Sampling must include fluids, melts, sediments, crust, and gases from the entire subduction zone, the subducting plate to the back-arc zone. Theoretical and experimental studies are an essential addition to these geochemical and geophysical programs. Existing studies have related only to isolated aspects of mass fluxes at convergent margins. However, the results of these studies can be used in conjunction with samples from existing and planned Deep Sea Drilling Project (DSDP) and Ocean Drilling Program (ODP) holes drilled offshore of various trenches to formulate an integrated, multidisciplinary plan for studying these complex systems.

Ocean Basin Sediments

Marine sediments provide important records for important Earth processes. For example, marine sediments furnish a history of regional and global volcanic activity, a record of the long-term changes in Earth's magnetic field, and a tool for studying large-scale tectonic processes, such as continental accretion and rifting. An active area of research is the study of past global climates.

Evidence of global environmental change comes from the paleoclimate record, which is the only long-term record available. The paleoceanographic record provides information necessary to understand environmental changes (see "Directions for Marine Geochemistry"). The paleoceanographic record also places observations of the present ocean in a historical context of long-term environmental variability. It affords a unique opportunity to test our understanding of the climate system as represented by numerical models of the ocean-atmosphere system. If models of the present ocean are capable of hindcasting oceanographic conditions, then we should have more confidence in their predictive capability. Whereas satellites provide a global means to observe the ocean surface, ocean sediments and the proxy indicators of oceanographic processes defined by paleoceanographers provide a global array of sensors to monitor processes within the ocean over long time scales. However, we must understand the relationships between these proxy data and modern processes to use this global information effectively.

With sampling resolution ranging from annual to interannual to millions of years, study of the marine sediment record allows the study of past climates on a wide range of time scales. Six specific research themes need to be addressed to improve significantly our understanding of global climate change and its effects; they are described briefly below.

Short-Term Spatial and Temporal Variability in the Climate System

It is desirable to characterize the natural variability in the climate system on annual and interannual time scales over spans of thousands of years. With this information, the significance of instrumentally observed climate changes can be assessed and the variations related to human influences. Marine sediment records already analyzed provide a qualitative, although spatially limited, picture of variability over the past 1,000 years, which appear to have contained several intervals of colder than normal climate (e.g., the Little Ice Age) as well as possible warmer time intervals (e.g., the Medieval Warm Period). Information on the magnitude and frequency of short-term variability should make possible a substantially improved assessment of the degree to which present trends are associated with increasing greenhouse gas influences.

Geological Record of the Carbon System

The objective is to identify and understand the role of the carbon system in past climatic change by isolating the response of global climate to natural changes in atmospheric carbon dioxide and other greenhouse gases. Studies of ice cores spanning the past 160,000 years now provide direct evidence that atmospheric carbon dioxide has changed over a large range (180 to 300 parts per million) during this period. Geologic evidence for the more remote geologic past suggests that atmospheric carbon dioxide may have been as high as four to eight times its present level. Knowledge of these large atmospheric carbon dioxide changes on geologic time scales presents an opportunity to understand global climate change and to test model estimates. For the period of ice core records, there is a well-documented forcing of climate changes by the distribution of solar radiation owing to Earth's orbital parameters. The response of climate to these forcing functions can be quantified through study of the geologic records. Thus the task is to document the interaction between the oceanic and terrestrial carbon cycle and atmospheric carbon dioxide as well as the pattern of climate change.

Instabilities in Ocean-Atmosphere Circulation in Earth History

Evidence strongly suggests that ocean circulation is sensitive to climate change, and changes in ocean circulation in turn influence the nature of the climate equilibrium (Kennett, 1977). The geologic record provides evidence for rapid, short-term transitions in deep-water circulation, associated changes in surface circulation and upwelling, climate changes during recent geologic history, and several abrupt reorganizations in ocean circulation over the past 60 million years. The ability to characterize the transitions in ocean circulation and to define, independently, the nature of the changes in the atmosphere will provide the means to describe case studies of the links between the ocean and the atmosphere.

Historical evidence confirms ocean response times on the order of decades, even for deep water, and a close link among climate and moisture fluxes, salinity, and deep-water circulation. Modeling studies indicate the potential for abrupt transitions between modes of deep-water circulation associated with little or no change in external forcing or with implied changes in surface moisture fluxes.

Episodes of Moderate to Extreme Warmth

Several intervals during the past 100 million years were significantly warmer than the present. Proxy evidence and the results of preliminary global circulation model sensitivity studies suggest carbon dioxide levels significantly higher than today's as the likely explanation of the global warmth during most of these episodes.

Geological Record of Global Sea-Level Change

The geological record contains widespread stratigraphic evidence of sea-level rises and falls (Shackelton, 1987), but further studies are needed to determine the magnitudes, rates, and causes of sea-level changes. Before the mechanisms of sea-level change can be addressed meaningfully, better estimates of the magnitude and rates of sea-level changes during the preglacial and glacial past must be obtained. It is especially important to determine upper limits to the rate at which the sea level can rise. The constraints provided by better-magnitude estimates could eliminate several postulated causes of sea-level change and help focus on the most relevant possibilities.

Effects of a Changing Physical-Chemical State on the Evolution of Marine Life

The fossil record preserved in the ocean is the best source of information on evolutionary dynamics, as well as a powerful tool in forecasting the biological effects of global change. It provides an exceptionally detailed picture of the distributions of fossil species and the climatic conditions in which they lived. Many extinct marine fossils have living counterparts that can be studied for knowledge of the ecology and genetics of extinct species.

Were organisms able to acclimate to the new environmental conditions and, if so, how? The biological effects of extreme shocks to the biosphere can also be examined, such as that imparted by the asteroid impact with Earth at the close of the Cretaceous. These events permit us to evaluate both how organisms respond to the threat of extinction and how survivors set about repopulating the vacated environment. Many of these events can also be studied in the terrestrial fossil record. However, the higher resolution of the oceanic record permits a far more complete analysis of the forces that underlie evolutionary processes on a global scale than can ever be accomplished by using terrestrial organisms. This resolution permits us to evaluate the role of climate change as a driving force behind the production of new species, the extinction of existing species, and geographic shifts in populations. We can forecast the biological consequences of human-caused changes in the environment by examining similar events in the fossil record.

Research Approaches

Study of oceanic crust and sediments has been aided over the past 30 years by a number of new techniques, whose application will continue to yield information about the seafloor, a region of Earth that still remains largely unknown. Rock and sediment cores obtained through the DSDP/ODP have provided glimpses of the structure, composition, and the processes that formed these materials. Continued systematic drilling will be required to obtain a complete picture of the structure of the ocean crust and particularly the chemical composition and hydrothermal alteration processes. Drilling also allows geophysical and geochemical instruments to be placed within the drill holes to measure temperature, chemical fluxes, crustal strain, and other variables important for understanding geological, geochemical, and geophysical processes.

The technique of seismic tomography began when it was learned that by studying the propagation of seismic waves, both in terms of speed and path, through the Earth, features of Earth's structure could be discerned. Later, explosives and noise generated by "air guns" were used to generate sound that can be transmitted some distance into the seafloor and reflected back to acoustic receivers. These techniques have been used to gain a more detailed picture of the upper seafloor, particularly the sediment layer overlying the crust. The newest seismic technique has been to drill holes deep into the seafloor, placing acoustic sources in some holes and receivers in others, to produce a horizontal seismic tomograph of the intervening sediments and crust.

These methods provide a snapshot of structure and composition from which processes and fluxes may be inferred. As with other oceanographic disciplines, the importance of time-series observations for observing dynamic processes is critical. Scientists that study marine geology and geophysics will increasingly use time-series measurements of changing features, through repeat cruises, rapid-response measurement techniques, and particularly sensors moored on the seafloor. The area of fluxes is one in which chemical oceanography and marine geology and geophysics interact, because measurement of benthic chemical fluxes is important for both fields. Finally, the concept of "seafloor observatories" is being implemented through the RIDGE program and through research sponsored by the Office of Naval Research. These "laboratories" are actually areas of the seafloor where repeated intensive observations are made. For example, ONR has designated sites on the Mid-Atlantic Ridge (a slow-spreading ridge) and the East Pacific Rise (a fast-spreading ridge) as natural laboratories for comparative studies.

DIRECTIONS FOR BIOLOGICAL OCEANOGRAPHY

Summary

In the next decade, biological oceanography will emphasize the effects of ecosystems on global cycles of important elements, such as carbon, nitrogen, and oxygen, and conversely the effects of global environmental changes on marine ecosystems. Of timely interest are climate change and population dynamics of marine organisms. In addition to climate change that may be accelerated by carbon dioxide and other greenhouse gas emissions, overfishing, eutrophication, introduced species, and other anthropogenic changes affect marine populations, although impacts vary regionally.

The complexity of biological systems and their variability in both time and space pose practical problems for designing programs and setting research priorities. Potentially important approaches include both studies focused on regions or times of the year with clearly distinguishable food-web structures and intensive examination of areas where geochemical measurements have identified inconsistencies or contradictions. There is also an urgent need to initiate and strengthen long time-series studies of the biology and chemistry of key oceanographic regimes. In addition, concerted effort must be applied to increasing understanding of the basic ecology, physiology, and molecular biology of key marine species.

For the foreseeable future, biological oceanographers will need ships to collect seawater, sediments, and organisms and to prepare and process samples at sea. Thus oceanographic vessels will remain the primary facility for advancing basic knowledge of marine ecosystems. However, the use of other technologies and approaches could lead to important breakthroughs. They include satellite and aircraft remote sensing; numerical modeling; molecular biological techniques; optical, acoustical, and sample collection instrumentation and in situ data acquisition systems, including bottom landers; and remotely operated vehicles.

Introduction

Studying marine communities is difficult without an understanding of their associated physical, geological, and chemical environments. It is likely that biological oceanographers will strengthen interdisciplinary collaboration in the 1990s to include more atmospheric chemists, meteorologists, sedimentologists, paleontologists, and other Earth scientists. At the same time, advances in biological oceanography will contribute information critical to studies in chemical oceanography and related disciplines. The ocean is a biochemical system, and the biotic and abiotic components of seawater coevolved, resulting in a distribution of elements in the world's ocean that is dictated by biological processes in the sunlit surface waters.

Biological oceanographers study the regulation of plant, animal, and bacterial production; the mechanisms affecting the way production is partitioned among trophic levels and individual species; and the dynamics of marine populations. They use various approaches to study these phenomena. Some biological oceanographers measure concentrations of carbon, nitrogen, calories, and other basic constituents of life and the rates at which they are transferred through the food web and to the seafloor as sinking particles. Others explore the physiology, behavior, genetic diversity, and abundance of individuals within populations and use this knowledge to develop conceptual models of the nature and regulation of marine communities. In recent years, molecular biology tools have contributed to these measurements. Research in the next 10 years will emphasize how biota affect global cycles of carbon, nitrogen, phosphorus, oxygen, and other key elements and, conversely, how climate and other ocean environment changes affect marine ecosystems. During the 1990s and beyond, studies of marine ecosystems are likely to be central to resolving controversies surrounding the key issues of global change.

There is general agreement that the ocean is a significant sink in the global carbon cycle (and related cycles of nitrogen, phosphorus, silicon, and other biologically important elements) and thus is an important modulator of the greenhouse effect caused by the buildup of atmospheric carbon dioxide. Carbon dioxide in the surface mixed layer of the ocean is generally within 30 percent of saturation, whereas it is supersaturated in the deep ocean by as much as 300 percent with respect to the present atmospheric carbon dioxide level. The concentration gradient is maintained by the "biological pump" in the surface waters, which through biological fixation, packaging, and transfer results in a net downward flow of carbon to the deep sea. The ocean is a carbon sink because some of the organic matter synthesized by organisms in the sunlit upper ocean (the euphotic zone) settles to the seafloor, and some small fraction of that reaching the seafloor is eventually buried in marine sediments, where it may remain for millions of years. Annual carbon burial in marine sediments is 0.5×10^{15}

grams globally making this process the largest biotic sink of the global carbon cycle (Moore and Bolin, 1986).

One of the major uncertainties in the models of the global carbon cycle is the role of marine organisms in the ocean carbon budget. During the spring bloom in the North Atlantic, the air-sea carbon dioxide flux is strongly controlled by biological activity. However, the comparative magnitude of the ocean and terrestrial sinks of carbon is in dispute (Tans et al., 1990), owing primarily to lack of knowledge about air-sea gas exchange rates, the variability of carbon dioxide saturation of surface waters, and the effects of food webs on the production, reoxidation, sedimentation, and burial of carbon.

The rate at which dissolved or particulate matter passes through the horizontal plane at any particular depth in the ocean is called vertical flux, whereas lateral flux refers to flux through a vertical plane. In the ocean, the vertical flux of organic material (as well as the lateral flux of organic material between estuaries and waters above continental shelves and between shelf and oceanic waters) and its burial rate in ocean sediments are not simple linear functions of primary production. The structures of marine food webs (the number and type of organisms at various feeding levels and the feeding relationships among the organisms) in the euphotic zone, in mid waters, deep waters, and at the seafloor are key variables affecting vertical and lateral fluxes of biologically important elements.

As indicated above, marine food webs affect global biogeochemical cycles, and marine populations, in turn, are affected by changes in global climate and human-induced changes in ocean environments. Some of the best examples of climate effects on marine organisms come from European fisheries, for which long time series exist for fish catch and abundance in relation to key physical and biological variables. An extraordinary event occurred during the 1960s in the North Sea, where the abundance of codlike fish exploded as the herring population declined. This major change probably occurred in response to a period of cooling that decreased the abundance of certain zooplankton species during the time of the year when young herring require zooplankton as food (Cushing, 1982). The impact of El Niño on South American anchoveta populations is another well-known example.

Human activities also affect marine populations, particularly in estuarine and coastal waters, although anthropogenic effects are difficult to distinguish from highly variable natural cycles. Of particular concern are the long-term effects of nutrient enrichment (eutrophication) resulting from altered land use, and waste disposal as a reason for the low oxygen concentrations on continental shelves. An additional concern is the deliberate or accidental transport of species from one ocean to another by shipping or other activities, leading to outbreaks of the introduced species. Documenting the causes and effects of changes in marine populations is difficult, but new techniques and approaches will make this research possible in the future.

Effects of the Food Web on Global Biogeochemical Cycles

Phytoplankton, macroalgae, and sea grasses use energy derived from sunlight to incorporate inorganic carbon, nitrogen, phosphorus, and other elements into organic molecules that are the building blocks of life and sources of energy for nonphotosynthetic organisms that consume these plants. Some bacteria also synthesize organic matter using chemical energy (chemosynthesis) rather than sunlight as a primary energy source. In the ocean, most organic carbon produced by photosynthesis and chemosynthesis is ingested by zooplankton and higher animals, oxidized for energy, and ultimately respired as carbon dioxide in the upper few hundred meters of the water column. At the same time, nutrient elements, such as nitrogen, phosphorus, and some trace metals, are recycled and reused by phytoplankton. A variable fraction of organic matter is not recycled in surface waters; instead, it settles out of the upper ocean layers, thereby contributing to vertical fluxes. Thus fluxes of carbon, nitrogen, oxygen, phosphorus, sulfur, and other biologically important elements are controlled by food-web processes. A major research theme for the 1990s will be to describe the effects and possible controls that food-web structure and function have on fluxes from the euphotic zone to middepth and deep waters, to the ocean sediments, and into the geological record.

Episodic Export of Material from the Surface

The simplest description of the effects of marine food webs on vertical flux involves only the size and species of phytoplankton and whether the phytoplankton sink before being ingested. For example, the spring diatom bloom in the North Atlantic is thought to sink without significant predation, whereas where cyanobacteria are the dominant primary producers, sinking of organic material from the upper ocean is largely mediated by food-web processes.

In the latter case, vertical flux is small, mainly because of the number of steps in the cyanobacteria-based food webs, which lead to more recycling. In either case, the importance of the downward fluxes to the biota is that food resources are no longer available to the community from which they exit but seem to fuel successively deeper communities.

Most particles in the ocean are small and sink slowly. Particles that account for most of the transfer of material to the seafloor are the rarer, large particles that have both high mass and high sinking rates. They include the fecal pellets produced by large zooplankton, large aggregates of detritus and plant debris (marine snow), and living organisms. Zooplankton can increase vertical flux by repackaging and concentrating organic matter from small, slowly sinking phytoplankton and microorganisms into fecal pellets and mucous feeding structures that sink much faster than individual particles. Sinking flux varies by an order of magnitude among food webs. Food webs dominated by large zooplankton consumers may export a much greater percentage of consumed primary production than food webs in which phytoplankton are initially consumed by smaller protozoans and zooplankton, owing to the relative sizes and sinking rates of fecal pellets. Episodic zooplankton swarms could dominate the long-term average export of organic matter from surface ocean communities, but such swarms are often missed by short-term studies.

The activities of marine animals in breaking apart and consuming large aggregates on their way to the seafloor may also be significant, and as yet poorly quantified, factors in controlling particle flux. Many of these particles are consumed by animals as they sink and are converted into smaller fecal pellets, new animal growth, respired carbon dioxide, and dissolved organic matter.

Dissolved Organic Material

The measurement of dissolved organic material (DOM) is also of great interest to biological oceanographers and is an area of overlap between the disciplines of biological and chemical oceanography. The size, average age, and biological availability of the DOM pool are controversial, but the pool could be significant in global fluxes of carbon, nitrogen, and other biologically important elements. Furthermore, a major unresolved question is the degree to which DOM provides nutrition for the ubiquitous microbial community, which may use organic carbon at 10 to 40 percent of the rate at which phytoplankton use it.

Benthos

Deep-sea benthic organisms receive a slow nonseasonal rain of fecal pellets and dead organisms. Recent studies in the North Atlantic show that additional large pulses of organic particles arrive at the bottom within weeks to months following the spring phytoplankton bloom, probably accelerated by formation of marine snow particles. A complementary study in the same general area indicated that benthic organisms grow faster than previously believed, with maximum growth rates following the annual deposition of phytoplankton detritus from the spring bloom (Lampitt, 1990). An open question is the extent to which benthic organisms rely on these episodes of rich input. Certain large animals may metabolically cache food resources.

Ocean Margins

The role of coastal areas in global ocean carbon and nutrient cycles is controversial. Several issues remain, such as the percentage of seasonal and annual coastal production that is exported to the deep sea, the percentage of global productivity that takes place in the coastal ocean, and the extent to which the coastal ocean functions as a net carbon sink because of the massive inputs of nutrients. Interdisciplinary studies will be required to resolve the controversies regarding lateral exchanges between estuaries and the coastal ocean, and between coastal and deeper waters. This point is developed further in "Directions for Coastal Ocean Processes."

Biology of Hydrothermal Vent and Hydrocarbon Seep Habitats

Most oceanic food webs are based on photosynthetic productivity occurring in the upper regions of the ocean. A little more than a decade ago, it was discovered that dense bacterial and animal communities, which rely largely on in situ chemosynthetic activity, thrive at deep-sea hydrothermal vents and at hydrocarbon seep zones. Carbon fixation in these habitats is driven by highly reduced substances, such as hydrogen sulfide, that are exploited by both free-living bacteria and bacteria living within animal tissues.

The role of deep-sea hydrothermal vent systems in generating and dispersing fixed carbon is an area of active study. Although it is unlikely that carbon fixation at the hydrothermal vents is a major factor in total global carbon fixation, carbon fixation in the deep sea deserves study because of its uniqueness. Characterization of the global extent of these systems, the rates at which their free-living and symbiotic bacteria fix carbon dioxide, and the extent to which organic materials at vents are distributed to other regions of the oceans will be key areas of research for the next decade. Beyond understanding the biogeochemical role of these communities, studies of vent communities will give insight into the evolution and functioning of nutritious and detoxifying mutualism among organisms. Support for this work has a broad international base, such as through the RIDGE program, which supports multidisciplinary investigations of the biology, geochemistry, and geophysics of mid-ocean ridge-crest systems.

Study of these diverse ecosystems in which chemosynthetic processes replace or complement photosynthetic productivity is necessary to understand the complex nature of marine food webs and the full suite of exchanges and transformations that constitute the global carbon cycle.

Effects of Climate Change on Populations of Marine Organisms

The characteristics of a region that determine its suitability for any given organism include not only the availability of food and the abundance of predators but also the dynamic physical features (mixing and circulation) of the local environment that influence the success of recruitment, efficiency of feeding, and susceptibility of organisms to predation. Global change could affect oceanic animal populations by changing physical processes of significance to the planktonic organisms. At present, it is not possible to predict definitively the impacts of global change on the physical parameters of the ocean and the atmosphere. However, the effects of climate change can be partly anticipated by examining similar effects on shorter time scales, such as seasonal freshwater pulses, El Niños, and other infrequent oceanographic phenomena. Three examples illustrate how global climate change could affect the physical features and processes of the sea that influence the abundance, distribution, and production of marine planktonic animals.

High-latitude marine ecosystems may be more susceptible to global change than low-latitude marine ecosystems. If precipitation patterns change as estimated and global warming triggers the rapid melting of previously persistent ice fields and the retreat of glaciers, the volume of fresh water that enters polar waters (e.g., the Gulf of Alaska) is likely to increase substantially. The input of fresh water can be critical to coastal currents, as seen by the effects of the Mississippi and St. Lawrence rivers. Changes in precipitation (temporal and spatial) and the amount of ice melt could shift the direction and change the magnitude of coastal currents. Such physical changes will affect fish populations by affecting transport of eggs and larvae.

The effects of El Niño events on eastern boundary current ecosystems in the Pacific Ocean could serve as a model of the possible effects of global warming (McGowan, 1990) in terms of decreased primary and secondary production. In addition, an increased temperature differential between land and ocean could enhance coastal winds and hence wind-induced transport of surface water away from the shore, reducing the reproductive success of species that spawn offshore but rely on coastal habitats later in their life cycles. Stronger winds would also increase turbulence in the surface mixed layer, dispersing patches of planktonic food, and thereby making the food less available for fish.

The third example involves the impacts of a changing sea level. If sea level rises at a rate of 1 to 3 millimeters per year over the coming 50 to 100 years, profound impacts on nearshore habitats would result. In areas with broad, flat coastal plains, the width of the inner continental shelf may be expected to increase greatly. This change would wipe out many coastal habitats. In addition, distribution of the wave energy over a wider continental shelf may substantially modify the transport of planktonic organisms to shore, affecting the success of larval recruitment and the transition of organisms from larval to juvenile stages.

Other Anthropogenic Influences

Other human-induced environmental changes also affect marine populations, although they vary regionally and their extent is disputed. For example, McGowan (1990) reported no detectable change of pelagic species or of ecosystem structure in the California Current ecosystem despite extensive harvesting (fishing) of top predators and vastly increased inputs of pollutants. In contrast, the Baltic Sea ecosystem has changed significantly in the past 50 years in response to eutrophication (Kullenberg, 1986).

The incidence of unusual, and sometimes harmful, phytoplankton blooms is increasing in coastal waters around the world. The evidence is particularly compelling in European and Japanese waters, where long-term water quality monitoring programs exist. The precise causes of any bloom event are difficult to ascertain, but there is increasing evidence that unusual phytoplankton blooms are related to changes in nitrogen-silicon ratios caused by eutrophication. The food-web consequences of the global epidemic of noxious phytoplankton blooms could be severe in some areas.

Fishing activity also changes the structure of marine ecosystems, although the effects of overfishing are often difficult to resolve from long-period cycles in organism abundance. Overfishing of the Georges Bank off the northeastern United States has changed the composition of fish species. From the mid-1960s until the early 1970s, herring and haddock declined by about a factor of 10 due to severe fishing pressure. At the same time, squid, dogfish, and sand lance increased, probably because they filled the ecological niches of depleted haddock and herring stocks (Sissenwine, 1986).

Long-term studies of some coastal benthic communities suggest that they have changed significantly owing to eutrophication. A 20-year time series of benthic species abundance data at a station in Puget Sound suggests that eutrophication may be causing shifts in the dominant species as well as increasing the magnitude of population fluctuations (Nichols, 1985). Yet, even with decadal time series, cause and effect are difficult to ascribe unambiguously, in part because anthropogenic effects are difficult to distinguish from natural changes.

Research Strategies

The complexity of biological systems and their variability in time and space pose practical problems for designing programs and setting research priorities. Research based on the theme that food-web variability controls variability in fluxes of biologically important elements in the global ocean could take many forms; efforts must then focus on a subset of key questions and approaches.

One possibility is to take a comparative approach and focus studies on regions or times of the year with clearly distinguishable food-web structures, and to examine processes in the euphotic zone and in deeper waters. A second possible strategy is to plan biological studies to resolve seeming inconsistencies or contradictions obtained from geochemical measurements and models. For example, recent interest in vertical fluxes in the North Atlantic (Altabet, 1989) were inspired in part by geochemical studies indicating that conventional views of productivity and particle flux in nutrient-poor waters were inconsistent with geochemical data.

A third possible strategy is to study regions of the ocean that are anomalous vis-à-vis standard paradigms. What controls productivity in nutrient-rich areas of the sea, such as the subarctic Pacific and the equatorial Pacific? If this question could be answered unequivocally, it would indicate significant progress toward a general understanding of oceanic productivity.

In support of all three strategies is continued research on the ecology, physiology, and molecular biology of representative species from specific oceanographic regimes. Without an understanding of the basic biology of individual organisms, one cannot hope to understand how the marine food web works or to predict how the ecosystem will respond to change.

Technologies and Approaches for the 1990s

The pace of scientific progress is often closely coupled with the development and application of new technologies. Several technologies and approaches will aid the study of marine ecosystems in the 1990s and could lead to important breakthroughs.

Satellite Remote Sensing

By the mid-1990s, three variables will be measured simultaneously by satellite for routinely characterizing ecosystems and related environmental factors. The three variables are sea surface temperature; sea surface and near-surface ocean color to determine chlorophyll and water clarity; and sea surface wind fields for estimating rates of vertical mixing, air-sea gas exchange rates, and other wind-related processes, such as the seasonal changes in the depth of the surface ocean mixed layer.

Numerical Modeling

Two developments in modeling should make significant contributions to ecosystem studies in the 1990s. First, models are being developed that can be used to help form hypotheses regarding the role of oceanic biota in global nutrient budgets. These models ultimately will merge basic mathematical descriptions of biogeochemical cycles with general circulation models and, from given starting conditions, will attempt to predict the evolution of fluxes over time. Global models will be particularly useful conceptual tools as techniques are developed to incorporate tracer distributions and satellite data into modeling procedures. For studies of marine populations, models based on individual organisms show promise because they allow treatment of biological variability at the species level.

Molecular Biology

Marine organisms in general have not been studied as extensively as their terrestrial counterparts; relatively little is known about the biota of the world ocean. Barriers to rapid advancements in biological oceanography include the inability of conventional technology to distinguish rapidly among marine taxa and to resolve important questions related to marine community structure, flow dynamics, and their interrelationships. Similarly, advances in marine biology and biological oceanography are limited by the paucity of fundamental knowledge of the genetics, molecular biology, biochemistry, and physiology of marine organisms.

A new suite of elegant and sophisticated technologies and instruments for molecular biology has been developed in the past two decades that could greatly facilitate studies of marine organisms. The technologies of molecular genetics are now applicable to ocean science. These technologies, which allow one to manipulate and probe the most fundamental life processes in ways that were not previously imagined, will revolutionize knowledge of the processes and mechanisms that regulate population, species, and community structures in ocean ecosystems.

In general, molecular biology will aid the study of marine ecosystems in two ways: it will help to determine the physiological state of organisms, and it will help to identify and characterize the genetic structure of marine populations. Work in these areas will help both to identify the causes of biological variability in the ocean and to understand the implications of this variability for the stability and ecological balance of human-impacted ecosystems. For example, these techniques were used to discover archaebacteria and prochlorophyte phytoplankton, to study the role of marine viruses, to determine the diversity of marine bacteria, and to study the enzyme activity of marine organisms. Research in these areas is in its infancy, and new techniques of molecular biology will undoubtedly continue to play an important role in future research.

Acoustics

Sound is an extremely useful tool in biological and fisheries oceanography. The scattering of sound by organisms at many different trophic levels can be used for a variety of purposes. Schools of fish and patches of plankton can be located and tracked acoustically. It may eventually be possible to distinguish living from nonliving scatterers and to identify the biological scatterers by species. It may soon be possible to estimate biomass acoustically as a function of trophic level in the ocean. Sound scattering has been used commercially since the 1930s to locate fish schools, but only recently have multifrequency systems been available for quantitative study of animal plankton. The Multi-Frequency Acoustic Profiling System is capable of profiling zooplankton in the size range of 0.2 to 10 millimeters. General application of acoustical technology will require the development of inexpensive equipment and techniques to analyze and use the large volumes of data generated.

Bio-optics

Fluorometers, transmissometers, and spectroradiometers are used to measure phytoplankton populations, the turbidity of the water column, and the amount and wavelength of light that penetrate beneath the ocean surface at a given site. Correlating site measurements with measurements from satellite ocean color sensors provides the means to extrapolate phytoplankton measurements to a global scale. Mooring optical instruments together with current meters and temperature and salinity sensors provides a technique for collecting long (months) and highly resolved (minutes to hours) time-series measurements, permitting biological oceanographers to study what physical factors control phytoplankton populations. Moorings contribute data on variation over time and depth, whereas satellite sensors provide information on variation over the global ocean surface. Flow cytometry is another optically based technology that is extremely useful for characterizing the size and pigment composition of phytoplankton and bacteria and for sorting populations based on these and other criteria.

Imaging for Organism Enumeration

New techniques for imaging organisms in situ, now available, show promise for widespread application in the 1990s. These technologies include schlieren video systems and holography, which have been used in the laboratory to study zooplankton feeding behavior. When successfully applied in situ, three-dimensional analyses of individual organisms and their spatial relations will be possible on scales sufficient to resolve the behavior of individual organisms.

Time Series

The concept of acquiring long time series of key ecosystem variables at important locations in the global ocean is certainly not new. Yet, with the possible exception of tide-gauge stations, the routine collection of temperature data by commercial ships, and a few simple physical measurements, time-series measurement programs are rare. A notable example of long time-series biological measurements is the Continuous Plankton Recorder Surveys of marine plankton in the North Atlantic Ocean. Ongoing programs measuring biological variables (e.g., the California Cooperative Oceanic Fisheries Investigation) are generally poorly funded. Virtually all recent planning reports stress the importance of long time series to resolve key global change issues and to describe the fundamental attributes of marine ecosystem dynamics. Satellite sensors and bio-optical moorings provide one level of information, but many more in situ observations are needed. Federal agencies recently have recognized the importance of financially supporting long-term measurement programs. For example, NSF supports time series stations at Bermuda and Hawaii, and the National Oceanic and Atmospheric Administration and the Office of the Oceanographer of the Navy are planning the U.S. contribution to a global ocean monitoring system. These time-series stations could be considered the beginning of the biological portion of a global ocean observing system.

Ideas and technologies are in place to make significant progress during the next 10 years in determining the role of marine ecosystems in global ocean biogeochemical cycles and the effects of global change on marine ecosystems. Available technologies range from molecular probes to satellite sensors. The ideas cover a comparable range of scales, from hypotheses about predator-prey encounters at centimeter-length scales to those about interannual variability in global ocean primary production. During the past five years, biological oceanographers have conducted a number of workshops and issued a large number of planning documents and reports (Appendix III). The field is obviously not idea limited.

DIRECTIONS FOR COASTAL OCEAN PROCESSES

Summary

In the coming decade, coastal research will be more interdisciplinary than it is now. Understanding coastal processes will require interdisciplinary studies of biological, chemical, physical, and geological processes. There will probably be considerable progress on the exchange of materials across continental shelves and between the ocean and the atmosphere. For example, physical mechanisms for cross-shelf exchange and their interactions with nonphysical mechanisms should be thoroughly studied. By the end of the decade, estimations of air-sea fluxes of momentum, heat, and gases in a nonequilibrium sea state (typical for the coastal ocean) should be possible. Further, the understanding of biological, chemical, and geological processes that affect these fluxes should have advanced substantially. Considerable progress should have been achieved in understanding the complex inner shelf (water depths of 3 to 30 meters), where measurements are difficult to make and processes difficult to model because of the many more factors that influence the system, compared to the open ocean. In the same vein, a more predictive understanding of the flux of materials through estuaries will emerge as our knowledge of the interactions of biological, geochemical, and physical processes improves.

Cross-shelf exchange and its related biological, chemical, geological, and meteorological components will be an active research area. So too will ocean fronts and their implications for biology, chemistry, and meteorology. Fronts are distinct boundaries between water masses and are nearly ubiquitous in the coastal ocean. The mechanisms that create them offer their own sampling and modeling difficulties.

Other areas of research are the global implications of coastal nutrient, carbon, and trace metal cycles and the study of ecosystem structures, which affect chemical cycles. Because of the diversity of coastal ocean regions, significant progress in understanding coastal processes may be achieved in some areas, but it is unlikely that results from all coastal regions can be integrated within the decade. Emphasis on toxic algal blooms, ecosystem structure changes, the invasion or dominance of nuisance species, and other human-induced biological effects may well increase.

In terms of facilities, it is likely that the demand for research aircraft will increase over the decade. The deployment of moored instrument systems will probably increase, especially with newly developed sensors, such as complete meteorological and wave measurements and underwater sensors capable of measuring biologically, chemically, and geologically relevant variables. Such comprehensive packages already exist or are under development, but their use will become more routine as the decade progresses. The need to study small-scale dynamic features, such as fronts on the continental shelf and wind-forced mixing in estuaries, will create a greater demand for towed devices that can sample rapidly and repeatedly in three dimensions. These too are becoming available but will be used more frequently in the future. Similarly, there will probably be a need for new sediment samplers that can also measure near-bottom currents and sedimentary conditions. Advances in remote sensing of coastal areas will benefit the field.

Introduction

The coastal ocean lies at the junction between land and the open ocean and includes estuaries and embayments. By virtue of its location, it is a setting of unusual societal importance. Most of the world's population centers are located near the ocean, so that pollution, recreation, and shipping impact the coastal environment and are likewise affected by coastal processes. As the U.S. population continues to shift toward the ocean, these considerations will become increasingly important. Economically, the coastal ocean is also of great importance, for example, in terms of mineral (especially petroleum) exploitation, recreation, and fisheries. The conflicts among uses of the coastal region have heightened the public's awareness of the region—and of the need to study it in detail.

The coastal region is defined here as the portions of ocean and atmosphere extending seaward from the surf zone and the heads of tidal estuaries and overlaying the continental shelf, slope, and rise. Geologically, this region of the continental margin forms the transition between the thick continental crust and the thinner oceanic crust, both of which float on the underlying mantle. The continental shelf is essentially the submerged edge of the continental crust. Broadly speaking, continental margins are of two types (Figure 3-3). Those on the leading edges of crustal plate motions (often near trenches) tend to be characterized by narrow shelves, (e.g., the West Coast of the United States). Margins on the trailing, relatively inactive edges of continents tend to be characterized by broad, relatively flat shelves (e.g., the U.S. East

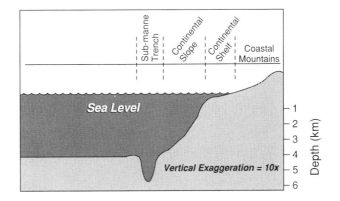

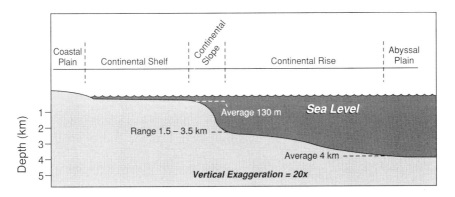

FIGURE 3-3 Typical profiles of two common types of continental margins. *Upper panel*: A collision margin typical of the Pacific coast of South America. The presence of a submarine trench, a narrow continental shelf, and a landward mountain range characterize this type of margin. *Lower panel*: A trailing-edge margin typical of much of the Atlantic Ocean. The presence of a continental rise, a broad continental shelf, and a coastal plain are characteristic of this type of margin.

Coast). The margins are often greatly modified by erosion and sediment deposition, processes that tend to carve out submarine canyons and fill in basins, respectively.

The following sections describe processes that make the coastal ocean unique and discuss some scientific issues that will be particularly important over the coming decade. Emphasis is on interdisciplinary aspects because it is likely that most important scientific and societal problems cannot be tackled successfully without a comprehensive approach.

Processes

The lateral boundaries and shelf-slope topography that characterize continental margins substantially determine the nature of coastal currents. For example, on a rotating planet, nearly steady currents are constrained from crossing isobaths (lines of constant depth). As a result, flow in the coastal ocean tends to parallel the coast, and exchange between waters over the continental shelf and the adjacent deep ocean is inhibited. Thus in many cases, distinct shelf water masses form, and the shelf represents a partially closed chemical and biological system. Fronts often mark the boundaries between these coastal and oceanic systems, and these fronts have their own important biological and atmospheric effects.

Wind-driven currents over continental shelves tend to be particularly energetic because the coastline interrupts water transport in the turbulent layer in the upper ocean. This interruption leads to a connection between surface winds and currents deeper in the water column. The resulting currents flowing alongshore below the turbulent surface layer dominate variability in most places over the continental shelves. Wind-driven currents are understood well enough that models are able to predict the speed and direction of coastal currents, as shown by the close agreement between observed and predicted currents shown in Figure 3-4.

Of broader importance to coastal ecosystems is the related onshore-offshore circulation, including the coastal upwelling of cold, nutrient-rich subsurface waters. Their temperature leads to the unusually cool, stable atmospheric conditions that characterize the U.S. West Coast during spring and summer. The upwelled nutrients fuel marine plant growth, leading to high biomass throughout the food web and some of the world's greatest fisheries, including those off the West Coast and off the coast of Peru. Upwelling can also intensify the transfer of organic materials from the surface to the seafloor in such areas. For example, off Peru, as much as one-half of the carbon fixed by phytoplankton production induced by upwelling may be deposited on the bottom. Upwelling in the coastal ocean can also be caused by factors other than wind. For example, upwelling of nutrient-rich water along the inshore edge of the Gulf Stream does much to stimulate productivity off the southeastern coast of the United States, as determined by chlorophyll measurements (Figure 3-5). Whatever its cause, upwelling contributes to the well-known high biological productivity of the coastal ocean (Figure 3-6). Estuaries and coastal embayments, on

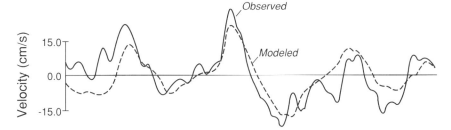

FIGURE 3-4 Observed (*solid line*) and modeled (*dashed line*) alongshore currents from over the continental shelf off Oregon, summer 1978. Positive velocity denotes northward flow. After Battisti and Hickey (1984).

the other hand, owe their high biological productivity to nutrient inputs from the land and density-driven internal circulation that serves to retain and enhance the recycling of these nutrients.

Sea ice is important in controlling air-sea fluxes in coastal regions when it forms there. Ice cover decreases heat, moisture, and gas fluxes and modifies momentum fluxes. During the formation of ice, salt is excluded, creating saltier adjacent water. These dense water masses can sink, impacting an entire basin through thermohaline circulation (see "Directions for Physical Oceanography"). Freshwater generated by ice melting stabilizes the water column, thus helping to prompt the spring phytoplankton bloom.

Tidal currents are sometimes enhanced over the continental shelves by physical resonances taking place in bays, such as in the well-known Gulf of Maine-Bay of Fundy example. Strong tidal currents intensify near-bottom mixing that can extend to the sea surface in shallow regions such as Georges Bank. This mixing and the resulting circulation enhance nutrient availability in the upper ocean, cause high primary productivity, enrich fisheries, and increase the transfer of organic material to underlying sediments. Energetic tidal currents can reinforce the many physical processes (including waves and wind-driven currents) that increase sediment resuspension and transport as well as the transport of chemicals that adhere to the particles.

Continental shelves are the transition zone between the land and the ocean and are thus particularly important in processes involving sediment and chemical fluxes. Freshwater outflows propel currents with distinct properties. Sediments from the land are

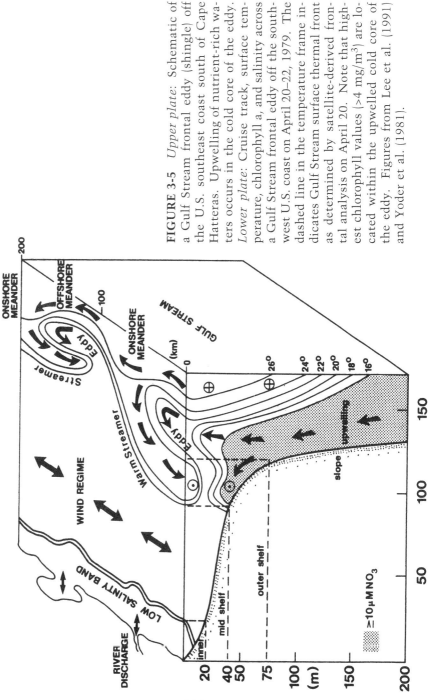

FIGURE 3-5 *Upper plate:* Schematic of a Gulf Stream frontal eddy (shingle) off the U.S. southeast coast south of Cape Hatteras. Upwelling of nutrient-rich waters occurs in the cold core of the eddy. *Lower plate:* Cruise track, surface temperature, chlorophyll a, and salinity across a Gulf Stream frontal eddy off the southwest U.S. coast on April 20–22, 1979. The dashed line in the temperature frame indicates Gulf Stream surface thermal front as determined by satellite-derived frontal analysis on April 20. Note that highest chlorophyll values (>4 mg/m³) are located within the upwelled cold core of the eddy. Figures from Lee et al. (1991) and Yoder et al. (1981).

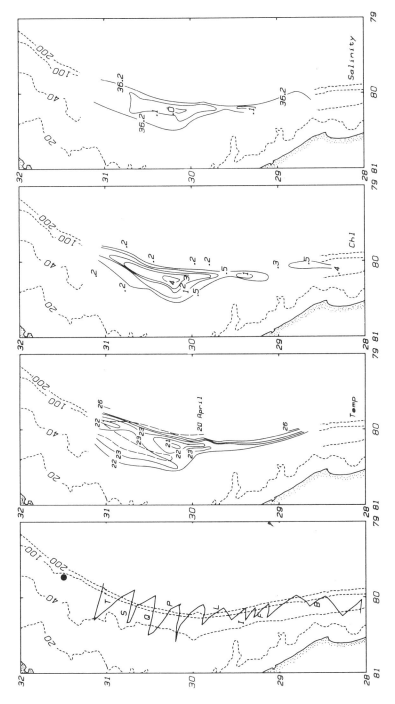

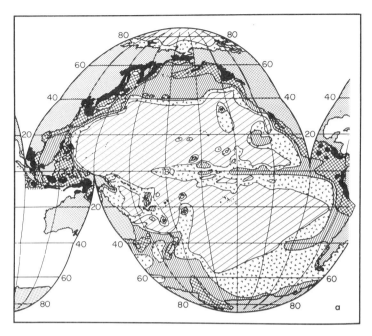

FIGURE 3-6 World ocean primary production according to Koblents-Mishke and coworkers on an equal area projection. Productivity categories are, from low to high, <36, 36-54, 54-90, 90-180, >180 gC/m²/yr. Note that most of the areas of high productivity are located on the ocean margins. From Berger (1988).

often deposited on the continental shelves, although they are sometimes transported to the slopes and deep ocean later. Sedimentary conditions on the shelf are far from static: numerous physical and biological processes can lead to reworking of the sediments and to their eventual transport to other locations. New evidence suggests that the shelf can be a source of particulates that accumulate within estuaries together with sediments delivered to the estuaries by rivers and shoreline erosion. Over geological time scales, the fates of sediments can vary widely with sea level; shelf processes can differ markedly, depending on how much of the shelf (or slope) is exposed above the sea surface. Coastal waters also receive chemicals and particulates weathered from continental rocks and transported to the ocean by rivers, groundwater, and winds. When these chemicals reach the coastal ocean, they are transformed or removed, so that although the properties of the estuarine waters may differ from those of the open ocean, shelf waters closely resemble open ocean water.

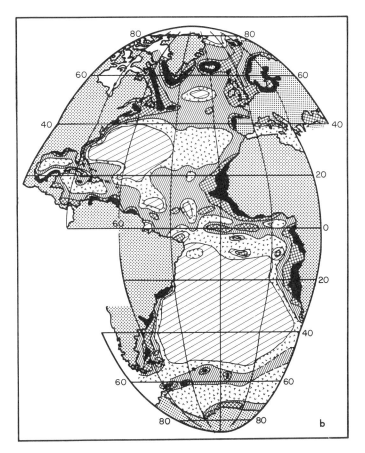

FIGURE 3-6 *Continued*

Physical processes on the scale of millimeters to kilometers have a major impact on the behavioral responses, feeding rates, interactions, and distributions of plankton, fish, and benthic invertebrates in the coastal ocean. For example, coastal fronts, island wakes, tidal flows, and vertical circulation cells are only a few of the many types of physical phenomena that can aggregate organisms or alter their behavior. Moreover, turbulence and small eddies on the scale of millimeters to meters partially determine the encounter rates of herbivores feeding on passive phytoplankton and bacteria and of predatory interactions among smaller pelagic organisms. An understanding of the effects of water movements on the behavior and distribution of organisms in the ocean will be one of the most challenging aspects of future research, particularly in coastal areas, where both physical processes and organisms are especially diverse and numerous.

Future Directions

The worldwide coastal ocean exhibits vast geographical diversity, depending on the size and openness of bays and estuaries; the width of the continental shelf; the proximity of strong oceanic currents; the strength of tides, winds, river runoff, and surface heat fluxes; and other characteristics. It is clearly impractical to explore fully the biological, chemical, geological, meteorological, and physical structure and variability of every estuary or shelf region of the United States, let alone of the world. One way to proceed is to identify the most significant physical-meteorological processes that to some extent act on all the world shelves and coastal waters. Each physical process and its effects on the biology, chemistry, and geology of the local area could then be studied in a prototypical environment (not limited to U.S. waters) where the process tends to predominate. The results of such interdisciplinary studies could be used to improve our modeling capabilities, enhancing our ability to model more typical shelves or estuaries where a combination of processes interacts. Although this approach is not a panacea, it can at least define the information needed to gain a desired level of understanding of a given coastal region. Within this broad approach to the coastal ocean, a number of important themes will be common to any detailed study of processes.

Air-Sea Interactions

The atmosphere is a major driving force of coastal ocean processes, through both its role in driving currents and its direct and indirect controls on biological and chemical processes. For example, wind-driven coastal upwelling can provide nutrients to the euphotic zone, leading to enhanced primary productivity, and atmospherically generated turbulence can increase predator-prey encounters among plankton (Rothschild and Osborn, 1988). Each of these biological processes results in distinct chemical transformations as well.

Present knowledge of atmospheric effects on the coastal ocean is limited to the effects of large-scale (500-kilometer) atmospheric features. This knowledge is useful for predicting alongshore currents or estimating the transport of dust particles from land to ocean (eolian deposition). Smaller scales in the wind field seem to be more important in determining cross-shelf currents, yet small-scale coastal winds are poorly observed and understood. Interaction of the atmosphere with the coastal ocean on these important scales of tens to hundreds of kilometers is not well-understood.

Air-sea fluxes of momentum and heat, for example, are not adequately characterized in present models, which do not take into account small-scale variability, directional offsets between the wind and waves, limited fetch, and limited water depth (which characterize the coastal environment; Geernaert, 1990). In addition, thermal fronts, which occur throughout the coastal ocean, greatly perturb the atmospheric layer directly above the sea surface and sometimes perturb weather systems. Further, the coastal topography helps to generate small-scale disturbances in the surface winds that can affect currents over the shelf. Air-sea fluxes of particles and chemicals, known to be important, must be a significant part of any study. Until we can quantify the air-sea momentum, heat, and chemical fluxes in this complex environment, we cannot understand the coastal ocean system as a whole.

Air-sea exchange is complex, but answers to the questions must be found. The atmosphere is the basic driving force of many coastal ocean processes. Ocean fluxes, especially heat fluxes, are critical to properties of the atmosphere. Air-sea exchanges that govern the effects of ocean and atmosphere on each other need to be quantified.

Cross-Margin Transport

The interaction of currents with bottom topography tends to isolate continental shelves from the rest of the ocean, although the strength of this isolation is significantly modulated by other processes. Even when the isolation is especially strong, shelf waters resemble the open ocean more than they resemble estuaries. It is difficult to identify which processes determine the cross-margin fluxes of water, particulates, chemicals, and organisms within estuaries, between estuaries and the shelf, on the shelf, and at the shelf-ocean boundary. The relative importance of such factors as wind-driven motions, frontal instabilities, turbulent boundary-layer transports, exchanges through submarine canyons, and the sinking of dense waters has not been evaluated. The difficulty is ultimately their episodic nature in terms of both location and time. Each has distinct effects on biological, chemical, and geological processes, so that interest in them is not limited to physical oceanographers.

Information on cross-margin transport is critical to all subdisciplines of coastal ocean science. Alongshore gradients of most characteristics tend to be small relative to cross-shelf gradients, and alongshore currents are relatively well understood. It is cross-shelf transport, or its absence, that shapes many distributions, such as those of sediments, that are of scientific interest. Estuarine and cross-shelf exchange is also of interest from a societal standpoint, for example, in determining the fate of riverine inputs of excess nutrients or pollutants. Thus it seems likely that estuarine and cross-shelf exchanges will be a central focus of future efforts in coastal ocean science.

Carbon Cycles

An important and controversial question in oceanography is, What is the role of the coastal ocean in global cycles of carbon, oxygen, nitrogen, and other significant elements? The coastal ocean occupies approximately 20 percent of total ocean area, yet accounts for approximately 50 percent of ocean primary production and approximately 50 percent of global ocean nitrate assimilation by phytoplankton (Walsh, 1991). Describing the mechanisms controlling cycling rates of essential elements has taken on new urgency because of the recently recognized potential for human alteration of global chemical cycles. Biological processes mediate the cycling of many elements and control the fate of numerous materials that enter the ocean. Constructing accurate models of biological controls and predicting their effect on the fate and transformation of dissolved substances and particles in the ocean are severely limited by our lack of understanding of the structure and function of marine ecosystems and their responses to physical and chemical processes. Elucidating these mechanisms is critical to understanding the coastal ocean because of its generally high productivity (and thus its processing capability), its substantial biological variability in space and time, and its role as a conduit between the continents and the deep ocean basins.

A major uncertainty in models of global change, including climate change, is the role of biological processes in mediating and controlling geochemical cycling of important elements. Most scientists agree that biological processes play a key role in the ocean carbon cycle and the cycle of nitrogen, oxygen, and related elements. However, the possible role of marine plants as a sink for carbon dioxide from human activities is highly controversial, and no generally acceptable model has been proposed to explain how the transfer of carbon from the ocean surface to the seafloor (the biological pump) should be working significantly faster now than before the Industrial Revolution. This is an important issue to be considered during the next decade. Understanding ocean margin food webs is of particular interest because they can be altered by eutrophication and other human activities.

In the ocean, the amount of organic material transferred vertically from the surface to the bottom and horizontally from estuaries to shelf waters to the deep ocean is not a simple linear function of primary production; nor are burial rates of organic matter in ocean sediments. The amount of material transported depends on the physical and chemical characteristics of the environment (e.g., rates and mechanisms of nutrient delivery) and on various largely unappreciated characteristics of the species composition and structure of marine food webs in the euphotic zone, deeper in the water column, and in and around the seafloor. Some biogeochemical cycling processes are summarized in Figure 3-7.

Particle Dynamics

Research in several areas needs to be initiated to improve our basic understanding of particle dynamics. Some of these areas have been mentioned, for example, the cross-shelf transport mechanisms and the use of narrow coastal margins with significant sediment inputs to model transport conditions during past times of lower sea level.

Among other research possibilities is the need to test the wide range of theoretical models for sediment transport that evolved in the past two decades. For example, models have been developed to describe the coupling between slowly varying currents and surface gravity waves and to predict resulting sediment transport.

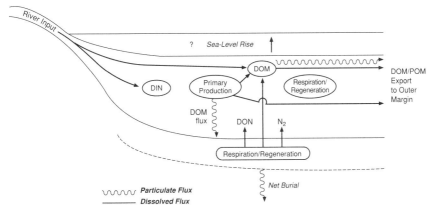

FIGURE 3-7 Schematic of some processes relevant to biogeochemical cycling on the inner continental margin; (DOM = dissolved organic material; DIN = dissolved inorganic nitrogen; DON = dissolved organic nitrogen; POM = particulate organic material). From Mantoura et al. (1991).

However, these models have received little or no testing in the laboratory or through field observations.

Muddy sediments are geologically relevant, and some research has been conducted on the transport of low concentrations of fine-grained sediments. Dense concentrations [>10 grams per liter (i.e., fluid muds)] are also observed in the marine environment, and their transport is poorly understood.

Carbonate sediments are widespread at low latitudes, but the effects of physical processes on their dispersal have not been thoroughly studied. Differences in the particle shapes and densities of carbonate sediments from more common sources make it difficult to extrapolate existing theory of sediment transport.

Theory regarding the formation of sediment layering within the seabed and its dependence on sediment transport and biological activity has evolved rapidly. Additional laboratory and field documentation is needed to link formative mechanisms and preserved strata.

The overall importance of the coastal ocean extends far beyond its relatively small areal extent. An environment of remarkably high biological productivity, this transition zone between land and open ocean is of considerable importance for recreation, waste disposal, and mineral exploitation. Such societal issues as pollution (in its many forms), bioremediation, waste disposal, and risk assessment cannot be addressed adequately until we make substantial advances in our basic understanding of the coastal ocean. A holistic approach to the coastal ocean system, blending marine meteorology with biological, chemical, geological, and physical oceanography, should enable us to progress sufficiently so that we will be better prepared to make the technical and policy decisions facing us over the next decades. Four issues of particular importance are air-sea interactions, cross-margin transport, carbon cycles, and particle dynamics. A balanced program would include studies focused on specific processes, long-term measurements, modeling, and instrumentation development. To take best advantage of the results of these studies, strong working relationships with the applied science communities need to be forged.

Coastal measurements will be an important part of a global ocean observing system because it is at the coasts that most countries, particularly developing nations, will make most of their measurements. Therefore, it is essential that the design of a GOOS include coastal measurements as a critical element of the system.

4

Human, Physical, and Fiscal Resources

HUMAN RESOURCES

Public and private institutions have developed an excellent graduate education system, yielding graduates employed in academia, government, and the private sector in the United States and abroad. The boundaries of oceanography are not well defined, and the field is characterized by many entry points from associated fields at various educational levels. Because of the diversity within the field and its relative youth as a separate science, a research oceanographer cannot simply be defined as one who holds a doctor's degree in ocean science. Many senior faculty in oceanography departments and institutions earned degrees in fields other than oceanography, and many scientists continue to enter ocean science from other fields. Nor can oceanographers be defined as those who perform basic research that is funded by the Division of Ocean Sciences of the National Science Foundation (NSF) or by the Office of Naval Research (ONR). Either definition misses many scientists whose primary activity is teaching, whose research is funded from other sources, or who are employed by federal agencies.

Ocean science will be characterized in the coming decade by a mixture of large multiple-investigator programs and individual investigations. The research will be only as good as the scientific talent that can be applied to the questions posed. Concern has developed regarding the potential shortage of Ph.D.s in science and engineering in the 1990s and beyond in terms of both number and quality. The oceanographic community has, however, questioned this assertion of a lack of qualified doctorates. This section discusses the demographics of oceanography and relates its characteristics to research needs.

In examining ocean science, the board asked eight specific questions:

• How many Ph.D.-level oceanographers are there, and at what rate has the number of Ph.D.-level ocean scientists changed over time?

• How many ocean science doctorates are produced annually?

• What is the present age profile of oceanographers in academia and the federal government, and has it changed over time?

• Has the field matured in terms of becoming a separate discipline?

• How has the percentage of women, minorities, and foreign nationals in the field changed over time?

• Has the field changed in terms of academic emphasis among the major subdisciplines [physical oceanography (P.O.), chemical oceanography (C.O) and marine chemistry (M.C.), marine geology and geophysics (MG and G), biological oceanography (B.O.) and marine biology (M.B.), and ocean engineering (O.E.)]?

• Has the balance of the field changed in terms of the relative size and importance of the major oceanographic institutions?

• How are research oceanographers supported? What is the ratio of institutional to federal salary support for the oceanography community as a whole?

Data Sources

Information was collected from a variety of sources. Data on the demographics of oceanography was obtained from biennial reports (1973 to 1989) issued by NSF, called *Characteristics of Doctoral Scientists and Engineers in the United States* (NSF, 1975; 1977; 1979; 1981; 1983; 1985; 1987; 1989; 1991). In addition, the Ocean Studies Board surveyed the major ocean science institutions and federal agencies (Appendixes IV and V). These two sources form the basis for much of the information presented. Additional information on faculty ages and number of Ph.D.s graduating was obtained from Joint Oceanographic Institutions, Inc.

(JOI). Data on ocean sciences grant recipient characteristics were obtained from NSF, and projected demands for Ph.D-level researchers were obtained from four major oceanographic research programs.

Results

National Science Foundation Surveys

Since 1973, NSF (through the NRC) has collected information on the employment and demographic characteristics of scientists and engineers with doctoral degrees in the United States. The NSF survey constituted a sample of the Ph.D. population, from which total population values were estimated. These estimates have substantial associated standard errors, so that statistical comparisons were not carried out. The number of oceanographers in all sectors of employment increased from 1,130 in 1973 to 2,460 in 1989 (Figure 4-1). From 1973 to 1981, the average annual rate of increase for academic oceanography was 4.7 percent; from 1981 to 1989, 4.0 percent. Oceanographers who consider teaching as

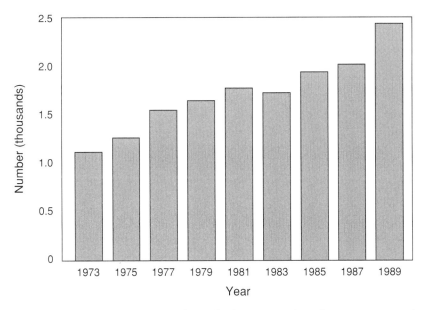

FIGURE 4-1 Change in number of Ph.D.s employed in oceanography over time (NSF data).

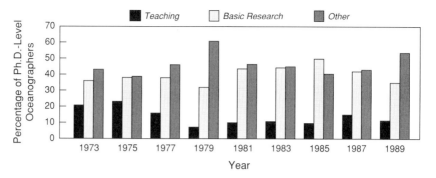

FIGURE 4-2 Primary work activity for Ph.D.s employed in oceanography (NSF data).

their primary work activity decreased from 21 percent in 1973 to 11 percent in 1989; the portion of oceanographers who consider basic research as their primary work activity fluctuated around 40 percent (Figure 4-2). Percentages in all employment sectors show no discernible trends over time (Figure 4-3). In 1989, most Ph.D.-level oceanographers—about 60 percent—were employed at educational institutions, including secondary schools, junior colleges, and four-year colleges. The federal government employed approxi-

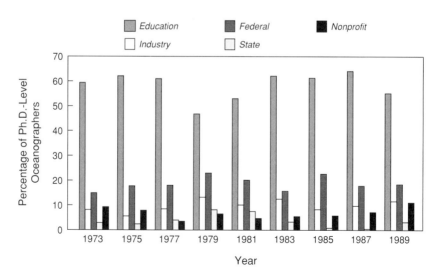

FIGURE 4-3 Employment sectors for Ph.D.s employed in oceanography (NSF data).

mately 20 percent of the nation's oceanographers; industry, about 10 percent; nonprofit organizations, 7 percent; and state governments, 4 percent. These percentages remained relatively stable over time.

The "maturity" of a discipline is the degree to which it is self-perpetuating and separate from other fields. Estimating the absolute maturity of a discipline is difficult, but examining changes in a number of indicators over time can show whether a field is advancing or declining. Two such indicators are the number of post-doctoral fellowships awarded and the ratio of faculty positions that are in the form of full professorships versus assistant professors. According to NSF data, the number of postdoctoral positions has increased, from an estimated 20 in 1973 to 84 in 1989 (Figure 4-4).

For new fields the ratio of full to assistant professors tends to increase over time because of the time required for faculty promotion and tenure, and the time universities need to establish tenured positions. For all science and engineering fields, the ratio has increased steadily over time, from 1.6 in 1973 to 2.4 in 1989 (Figure 4-5). The ratio for oceanography increased from 1.0 to 3.5 in the same period (Figure 4-5). The leap in the ratio in 1989 was due to a substantial increase in the number of full professors and a decrease in the number of assistant professors. The full to

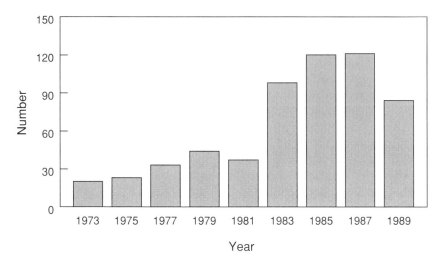

FIGURE 4-4 Postdoctoral fellows in oceanography (NSF data).

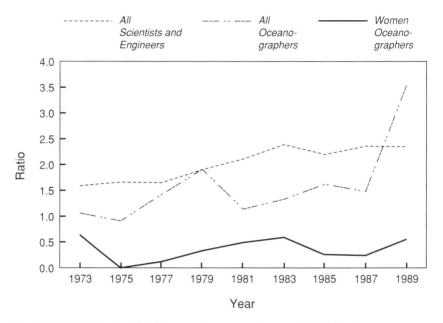

FIGURE 4-5   Ratio of full to assistant professors (NSF data).

assistant professor ratio is even lower for women, reflecting their relatively recent entrance into the field.

The proportion of the field made up of women increased from about 3 percent in 1973 to 11 percent in 1989 (Figure 4-6A). Minorities and foreign nationals practicing oceanography in the United States showed no significant trend from 1973 to 1989 (Figures 4-6B and C).

NSF data show that from 1973 to 1989, the median age of Ph.D. oceanographers shifted from the 35- to 39-year-old bracket to the 40- to 44-year-old bracket.

Ocean Studies Board Survey

Information on the potential supply of and demand for oceanographers is limited. Several attempts have been made to characterize the field over the past 20 years (NRC, 1970, 1972, 1981).

FIGURE 4-6   (See opposite page.) (A) Gender of employed oceanographers (NSF data). (B) Race of employed oceanographers (NSF data). (C) Nationality of employed oceanographers (NSF data).

A

B

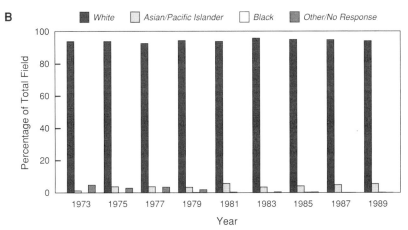

C

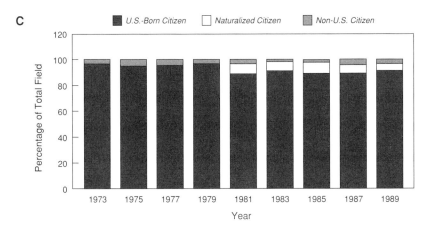

For this study, the Ocean Studies Board (OSB) sent questionnaires to 52 oceanographic institutions, research laboratories, and academic members of the Council on Ocean Affairs, and to 8 federal agencies to assess the supply and demand within the academic and federal sectors. Responses were received from 40 academic institutions, including all the large academic programs and research institutions, and from 7 federal agencies (Appendixes VI and VII). Of the 40 institutions employing oceanographers in 1990, only 29 had employed oceanographers in 1970.

Replies to the OSB questionnaire indicated that the number of academic oceanographers increased from 540 in 1970 to 1,674 in 1990 (Figure 4-7). These include both teaching faculty and research faculty. It should be noted that some of the growth in the 1980–1990 period for academic oceanographers was due to the inclusion of 378 faculty members from two newly created units, at the University of Hawaii (UH) and the University of Washington (UW), that had not been included in the totals before 1990. At the same time, the number of Ph.D. oceanographers in federal agencies rose from 148 to 516. The annual rates of increase (percent) were

	1970–1980	1980–1990
Academic	6.4	2.6 (without UW and UH)
		5.2 (with UW and UH)
Federal	9.9	3.1

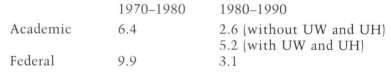

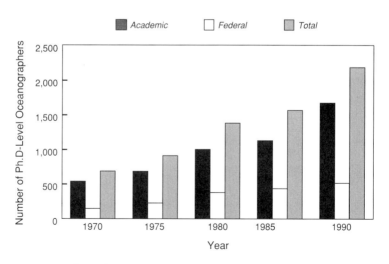

FIGURE 4-7 Ph.D.-level federal and academic oceanographers (OSB survey).

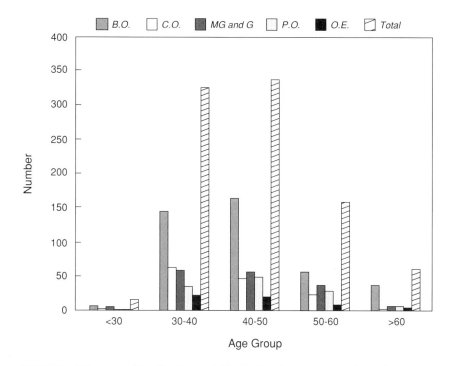

FIGURE 4-8 Age distribution of Ph.D.-level oceanographers in oceanographic institutions and universities (OSB survey).

Figures 4-8 and 4-9 show that for universities and government laboratories, respectively, the largest number of oceanographers in any age range falls in the 40- to 50-year-old category. The marked peak in the age distribution of federally employed oceanographers could reflect the establishment and expansion of federal oceanography programs in the 1970s.

The ratio of full to assistant professors in ocean sciences over the past 20 years has increased from 1.0 to 1.6 (Table 4-1). During roughly the same period, NSF data show an increase from 1.0 to 3.5. This reason for this discrepancy in unknown, although the large standard error in the NSF data makes comparisons difficult. Figure 4-10 shows the increase in Ph.D.-level staff by rank. The number of postdoctoral positions increased from 11 in 1970 to 111 in 1990, according to OSB data, compared with an increase from 20 in 1973 to 84 in 1989, according to NSF data.

Figure 4-11 shows changes in the number of Ph.D.-level oceanographers by discipline over time, as determined by the OSB sur-

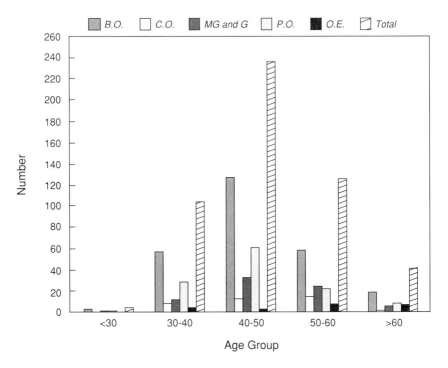

FIGURE 4-9 Age distribution of Ph.D.-level oceanographers employed by government agencies (OSB survey).

vey. The category that includes biological oceanography and marine biology continues to dominate numerically, reflecting the number of relatively small marine laboratories that focus on biological research. Except for a marked increase in ocean engineering, the relative ratios among the academic subdisciplines have not changed substantially over the past 20 years (Table 4-2). For

TABLE 4-1 Ratio of Full Professors to Assistant Professors in Oceanography, 1970–1990 (OSB survey)

Year	Ratio
1970	1.0
1975	1.2
1980	1.2
1985	1.6
1990	1.6

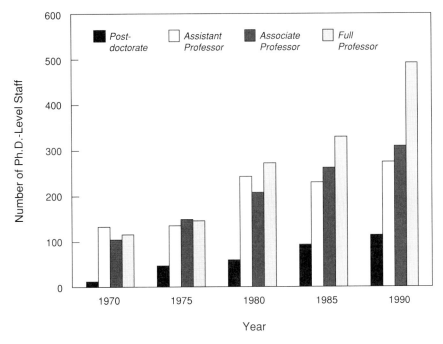

FIGURE 4-10 Rank of Ph.D.-level staff in academic institutions (OSB survey).

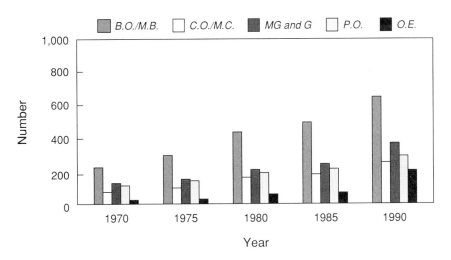

FIGURE 4-11 Change in number of Ph.D.-level oceanographers over time (OSB survey).

TABLE 4-2 Percentage of Ocean Scientists in Subdisciplines (OSB survey)

	Academic					Federal				
	1970	1975	1980	1985	1990	1970	1975	1980	1985	1990
Biological Oceanography/ Marine Biology	40.4	42.2	42.5	42.4	37.7	73.0	68.9	48.3	51.5	51.0
Chemical Oceanography/ Marine Chemistry	13.0	13.3	15.2	14.8	13.9	8.8	11.4	9.8	8.5	7.0
Marine Geology and Geophysics	22.6	21.3	19.7	20.2	20.8	3.4	2.6	13.5	14.0	13.4
Physical Oceanography	20.0	19.1	17.9	17.5	16.4	9.5	13.6	25.3	22.2	24.6
Ocean Engineering	4.1	4.1	4.7	5.0	11.2	4.7	3.1	4.2	6.9	5.8

federally employed oceanographers, the percentage of biologists has declined markedly, and the percentages of specialists in physical oceanography and marine geology and geophysics have increased (Table 4-2). The percentage of biologists in the federal government is considerably higher than in academia.

NSF, ONR, and JOI Institutional Data

The JOI members are 10 of the country's largest oceanographic institutions. In the most recent year for which data are available (fiscal year 1991), the JOI schools received 45 percent of the NSF Ocean Science Research Section funding and 42 percent of ONR funding (SE31 and SE32).

Figure 4-12 shows the percentage of faculty at JOI member institutions related to the total number of oceanography faculty, excluding data for the University of Washington and the University of Hawaii. In general, the percentage of the total oceanography faculty located in JOI institutions has not changed over time, although the percentage of marine engineers at JOI institutions may have increased, and biologists and chemists may have decreased (Figure 4-12). The JOI institutions, where the large ships are concentrated, still tend to dominate the field in the disciplines that require large ships, such as marine geology and geo-

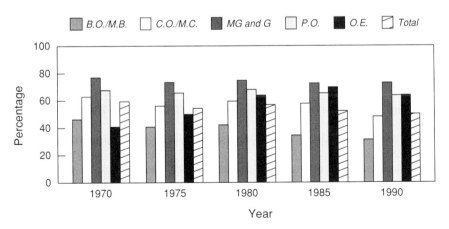

FIGURE 4-12 Percentage of oceanography staff at JOI institutions (OSB survey).

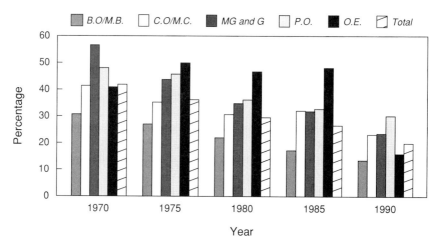

FIGURE 4-13   Percentage of oceanography staff at Scripps Institution of Oceanography and Woods Hole Oceanographic Institution (OSB survey).

physics and physical oceanography. This statement is less true for the biological sciences. If the same comparison is made for just Scripps Institution of Oceanography (SIO) and Woods Hole Oceanographic Institution (WHOI), the two largest oceanographic institutions, their combined dominance in terms of percentage of faculty has decreased steadily over the past 20 years (Figure 4-13), except in marine engineering. So although the percentage of total oceanography faculty at the two largest oceanographic institutions has decreased over the past two decades, the percentage of total oceanography faculty at the ten largest has remained about the same.

JOI provided information on its institutions' students, graduates, and faculty. The number of ocean science doctorates awarded annually at JOI institutions increased from 90 in 1970 to 126 in 1991 (Figure 4-14). The major change is the large increase in the number of women earning doctorates in the ocean sciences, up from 10 percent in 1980 to almost 30 percent in 1991. The number of foreign students earning doctorates is also about 30 percent; 2.5 percent of JOI students are underrepresented minorities.

The median age of oceanographers who received NSF grants increased from 40 in 1977 to 45 in 1990. The median age of JOI faculty was 44 years in 1990 (Figure 4-15).

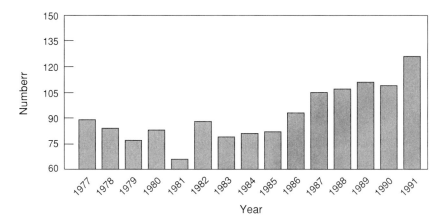

FIGURE 4-14 Number of Ph.D.s awarded annually at JOI institutions.

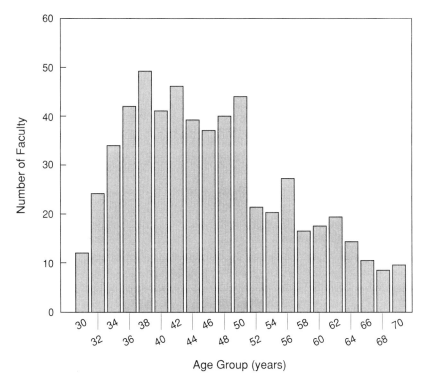

FIGURE 4-15 Age distribution of Ph.D.-level staff at JOI institutions in 1990.

Demand from the Major Programs

The extra demand required for the major ocean science initiatives planned for the 1990s is difficult to estimate. It is possible to estimate how many ocean scientists the major programs will require if the programs are funded at the projected levels, but the number of participants who are already in the field is an unknown. Because oceanography continues to attract scientists from physics, chemistry, biology, and geology, entry points into the field vary from the undergraduate to the postdoctoral level. Thus demand can often be met from associated fields. Nonetheless, it is of interest to estimate the impacts of four major oceanographic initiatives on human resources. The requirements of the programs were estimated by the individual program offices and represent a maximum level under a scenario of full funding and the assumption that the programs retain their original scopes and timetables.

The U.S. office for the World Ocean Circulation Experiment (WOCE) has estimated the work force that will be required to carry out its planned experiments for 1990–2000 (Table 4-3). The figures were extrapolated from NSF-funded project proposals. The total principal investigator (PI) and postdoctoral fellow labor-months estimated for the U.S. part of WOCE is 8,189, of a total of 28,507 (30 percent). The U.S. office for the Joint Global Ocean Flux Study (JGOFS) estimated a requirement of 14,000 labor-months for all categories from 1990 to 2000. If it is assumed that roughly

TABLE 4-3 Estimated Demand for Ph.D.s for Major Ocean Science Research Programs (1990–2000)

Program	Person-Years[a] All Ph.D.-Level Oceanographers	Postdoctoral Oceanographers
WOCE	1,000	320
JGOFS[b]	720	160
RIDGE[b]	200	40
GLOBEC	1,100	440
Total	3,020	960

[a]Assumes 6 person months per year for 10 years
[b]Assumes that JGOFS and RIDGE have the same ratio of PI and postdoctoral labor-months:total man-months (30%) as WOCE.

the same percentage of the total labor-months for WOCE scientists and postdocs should be valid for JGOFS, then JGOFS will require an estimated 4,300 labor-months (0.3 × 14,400) in this decade. The Ridge Inter-Disciplinary Global Experiment (RIDGE) office estimates it needs 4,000 labor-months over the 1990–2000 decade, and the Global Ocean Ecosystems Dynamics (GLOBEC) program has estimated 6,600 labor-months at the PI and postdoctoral levels. If 6 labor-months per labor-year are assumed, equal annual effort over the decade, and full program funding are assumed, approximately 300 Ph.D.s will be required to carry out WOCE, JGOFS, GLOBEC, and RIDGE. Of these, 100 will be at the postdoctoral level. If only 50 percent of the average oceanographer's labor-months are available for research, about 22 percent of the 1990 academic oceanographer pool would be needed for these four programs, if they are fully funded.

Answering Specific Questions

How many Ph.D.-level oceanographers are there and at what rate has the number of Ph.D.-level ocean scientists changed over time? According to the OSB survey, there were 1,674 academic oceanographers and 516 federal oceanographers in 1990. The NSF survey (1989) estimated 1,354 academic oceanographers, 453 federally employed oceanographers, and 653 Ph.D.-level oceanographers in other sectors.

The growth rate in the number of Ph.D.-level oceanographers slowed from the 1970s to the 1980s. Average annual growth rates for the pool of academic oceanographers decreased from 4.7 to 4.0 percent according to NSF surveys, and from 6.4 to 2.6 percent according to the OSB survey. The slowing of growth was even more evident for the federal government.

How many ocean science doctorates are produced annually? The JOI data show that approximately 126 oceanography Ph.D.s were awarded from JOI institutions in 1991, which is the largest number in any year for which data are available.

What is the present age profile of oceanographers in academia and the federal government, and has it changed over time? The OSB survey measured a median age in the 40- to 50-year-old bracket for both academic and federally employed oceanographers. The JOI faculty age distribution shows a median of approximately 44 years. The median age of the field has increased over the past 20 years from the 35- to 39-year-old bracket to the 40- to 44-year-old bracket, according to the NSF survey. In addition, the median age of NSF Ocean Sciences Division grantees increased from 40 years in 1977 to 45 years in 1990.

Has the field matured in terms of becoming a separate discipline? Over the past 20 years, the field has matured according to several measures. The expansion of postdoctoral positions shown by both the NSF and the OSB surveys, and the increase in the ratio of full to assistant professors are both indicators of the field's maturing. The significance of changes in the faculty ratio is uncertain, however, because the ratio for the combined science and engineering fields has also increased, and the 1989 jump in ratios for oceanographers is difficult to explain. The lag of female faculty behind the rest of the field may be because of the relatively recent entry of women into the field.

Has the participation of women, minorities, and foreign nationals changed over time? The percentage of women in the field of oceanography increased from 4 to 11 percent from 1973 to 1989, according to the NSF survey. At present, 30 percent of students at JOI institutions are women. The percentage of underrepresented minorities is low in both the population of employed oceanographers (7.7 percent) and the JOI student population (2.5 percent). The percentage of oceanographers working in the United States who are foreign nationals did not change dramatically from 1973 to 1989.

Has the field changed in terms of emphasis among the differing major subdisciplines (physical oceanography, chemical oceanography, marine geology and geophysics, biological oceanography and marine biology, and ocean engineering)? The relative balance of the number of scientists in the academic disciplines has changed little in the past 20 years. For federally employed scientists, fewer are biologists and more are specialists in physical oceanography and marine geology and geophysics now than in 1970.

Has the balance of the field changed in terms of the relative size and importance of the major oceanographic institutions? This analysis documents the fact that some decentralization of the field has occurred over the past 20 years in terms of where Ph.D-level scientists are employed. During and after World War II, Navy and NSF support led to the expansion of JOI institutions. In 1970, the faculty at SIO and WHOI constituted approximately 40 percent of the field. By 1990, the faculty at these two institutions comprised only about 25 percent of the total. The distribution of scientists at JOI institutions differed by subdiscipline, correlating with sciences that tend to require large ships, such as physical oceanography and marine geology and geophysics. In terms of financial support from NSF, JOI institutions received a relatively

TABLE 4-4 Support of Ocean Science Faculty at Academic Institutions

Faculty Position	Number of Months of Institutional Support	
	JOI Schools	Non-JOI Schools
Full professor/scientist	7.3 ± 1.0 (n = 8)	8.5 ± 0.5 (n = 23)
Associate professor/scientist	5.5 ± 0.9 (n = 8)	7.8 ± 0.6 (n = 23)
Assistant professor/scientist	4.9 ± 1.1 (n = 8)	7.7 ± 0.6 (n = 28)

NOTE: n = the number of institutions responding. It is assumed that each institutional response is the average of that institution's professionals.

constant 45 percent of NSF ocean science research funding between 1984 and 1989. JOI institutions received about 40 percent of ONR funding (SE31 and SE32).

How are research oceanographers supported? What is the ratio of institutional to federal salary support for the oceanography community as a whole? Oceanographers' salaries come primarily from grants and contracts. Academics from JOI institutions must raise a significantly greater proportion of their funding from external sources than other academics. The OSB survey shows that most of the oceanographic community, especially JOI schools, depends on noninstitutional research support (Table 4-4).

PHYSICAL RESOURCES

The wide variety of facilities used in institutions and consortia for ocean science—ships, submersibles, satellites, special platforms, and laboratories—depends on continual renewal to meet present and future needs. Global change research has given new impetus to satellite data systems and large-scale at-sea programs. Although oceanographers learned to use satellite data in the past decade, incorporating the increasing stream of data from new satellites and platforms will be a technological and managerial challenge.

Oceanographic Institutions

From its beginning, a mix of government, university, and private laboratories has conducted oceanographic research. The history of our ocean science institutions is characterized by three phases. Civilian marine science began in the late 1800s with the establishment of several marine biological laboratories concerned principally with coastal problems. The California Academy of Sciences (1853), the U.S. Fish and Wildlife Biological Laboratory at Woods Hole (1885), Hopkins Marine Station of Stanford University (1892), and the Hydrobiological Laboratory of the University of Wisconsin (1896) were notable among the early laboratories.

Between the turn of the century and the end of World War II, both the number of ocean science laboratories and the disciplinary range of their activities grew. During this period, Scripps Institution of Oceanography (1903), Friday Harbor Laboratories of the University of Washington (1904), Woods Hole Oceanographic Institution (1930), Narragansett Laboratory of the University of Rhode Island (1930), Bingham Oceanographic Foundation of the University of Southern California (1940), the Virginia Institute of Marine Science (1941), and the University of Miami Marine Laboratory (1943) were established. Several of these laboratories continued the thrust of activity in coastal marine biology, and many expanded into physical, chemical, and geological oceanography and increasingly carried out research in the open ocean.

World War II was a major turning point in oceanography. Research on ocean processes begun during the war continued afterwards as basic research programs supported by the newly created Office of Naval Research. Additional ships were added to the oceanographic fleet, and support for both research and ship operations was readily available. Under the Navy's leadership during the postwar period, growth in the number of ocean institutions and their scope of research accelerated. Thus from the late 1940s to the early 1950s, several laboratories, most of which would eventually engage in deep-ocean research, were established or expanded. Among the new institutions were the Chesapeake Bay Institution of the Johns Hopkins University (1948), Florida State University Oceanographic Institute (1949), the Department of Oceanography of Texas A&M University (1949), the University of Delaware Marine Laboratories (1951), the Department of Oceanography of the University of Washington (1951); the Department of Oceanography of Oregon State University (1958), and the University of Hawaii Institute of Geophysics (1959).

In the early years of marine science, there were no formal mechanisms for coordinating institutions' activities. The Joint Oceanographic Institutions for Deep Earth Sampling (JOIDES), an international advisory committee, was established in the late 1960s to provide formal advice to the Deep Sea Drilling Project (DSDP). JOIDES was a major initiative by the oceanographic community to develop a mechanism for international cooperative activities. Evolving from JOIDES was JOI, a formal not-for-profit corporation. JOI consists of 10 U.S. ocean science institutions that operate many of the large ships in the oceanographic fleet, employ a majority of U.S. academic ocean scientists, and receive a majority of the research funding. The JOI institutions are

> Scripps Institution of Oceanography, University of California
> Lamont-Doherty Geological Observatory, Columbia University
> School of Ocean and Earth Science and Technology, University of Hawaii
> Rosenstiel School of Marine and Atmospheric Sciences, University of Miami
> College of Oceanography, Oregon State University
> Graduate School of Oceanography, University of Rhode Island
> College of Geosciences and Maritime Studies, Texas A&M University
> Institute for Geophysics, University of Texas
> College of Ocean and Fisheries Sciences, University of Washington
> Woods Hole Oceanographic Institution

With the exception of Woods Hole, which has a joint education program with the Massachusetts Institution of Technology (MIT), each oceanographic program is an integral part of a major university.

Another cooperative organization of oceanographic institutions is the University-National Oceanographic Laboratory System (UNOLS), an association of ship operators and ship users that is discussed in more detail below. Because UNOLS provides access to facilities for scientists at institutions without ships, an increased number of universities can be involved in open ocean research. These universities may not have interests in all facets of oceanography, but they have significant strengths in certain areas. Examples of such universities are the Santa Cruz and Santa Barbara campuses of the University of California, Northwestern University, Massachusetts Institute of Technology, and Princeton University.

The institutions developed within and outside the government for the pursuit of an understanding of the ocean are diverse, much more so than in most other scientific fields. Oceanography is conducted by individuals working as faculty members in conventional academic Earth sciences departments supported by state and private endowment funds (e.g., MIT, Florida State University, the University of Michigan) in large research institutions operated by universities, but on a scale not common to academic institutions (e.g., Scripps Institution of Oceanography, Lamont-Doherty Geological Observatory), in independent, private nonuniversity organizations (e.g., Woods Hole Oceanographic Institution, Boothbay Harbor Laboratories, Monterey Bay Aquarium Research Institute); in government laboratories resembling the private laboratories in many ways (the National Oceanic and Atmospheric Administration Atlantic Oceanographic and Meteorological Laboratories and Pacific Marine Environmental Laboratories); and in Navy laboratories charged with specific military responsibilities.

This diversity is both a potential weakness and a strength. Oceanographers are generally more dependent on grant money than are scientists in other disciplines who receive a higher percentage of support from their universities. On one hand, this situation renders ocean science more vulnerable to government budget fluctuations. However, the institutions are adaptable to changes in the conduct of ocean science. Some institutions are expert in seagoing observations, some specialize in ocean engineering, some are focused on large-scale numerical modeling, and others are best known for their breadth. Together, they comprise the strongest marine research establishment in the world.

Most oceanography degrees are offered at the graduate level; however, an increasing number of institutions are now offering undergraduate degrees in oceanography. Integration of marine research facilities (often isolated from the campus) into the academic structure of the parent university is improving, and new oceanography programs have developed within a more traditional academic departmental structure. Perhaps this change can be considered an indicator of the maturing of oceanography as a recognized academic discipline.

Several new organizations of ocean science institutions have recently formed, such as the Council on Ocean Affairs (COA) and the National Association of Marine Laboratories, to promote interlaboratory cooperation. COA is an organization of approximately 50 academic oceanographic institutions that was founded by, and is administratively housed in, Joint Oceanographic Institutions, Inc.

Thus, with increased ease of access to the sea for faculty and students, the establishment of more oceanography activities in universities, and substantial support by some universities, oceanography is becoming an established academic discipline. Physical resource requirements to ensure that the levels of support, equipment, and access to the ocean are adequate to carry out the research needed in the next decade should be important principles as academic institutions and federal agencies develop new partnerships.

Ships

Even with new remote sensing techniques and autonomous vehicles, ships will continue to be the major platform for direct at-sea observations and measurements as well as for the calibration and verification of remote measurements. These tasks require a modern fleet of research vessels, a fleet whose composition and capabilities should be tailored to research objectives.

The federal oceanographic fleet is defined as the set of oceanographic vessels whose operations are funded by the federal government. The fleet is composed of more than 60 vessels operated by both federal agencies and academic institutions. The academic institutions coordinate their ship activities through UNOLS, which was formed in 1971 to support oceanographic research by coordinating and scheduling ships and equipment for their efficient use. UNOLS institutions operate and use vessels owned by the NSF, the Navy, and academic institutions. The UNOLS fleet, although not formally designated as a national facility, is recognized as a national asset vital to the needs of U.S. oceangoing scientists. Before the formation of UNOLS, each institution negotiated separately with the group of federal supporters. Ships were scheduled primarily for the exclusive use of the operating institution's scientists. UNOLS's consolidated scheduling of ships has improved efficiency and ensured availability of time at sea to all funded researchers. Its success has reduced the importance of each institution's operating its own research vessel and has allowed, from a national viewpoint, institutions without ships to develop strong marine programs with seagoing components.

UNOLS consists of 57 member institutions, of which 20 operate research vessels. The UNOLS fleet is composed of surface ships ranging in length, age, and origin; the submersible *Alvin*; and the floating instrument platform (FLIP) (Table 4-5). Some were built using capital provided by the federal government; others were built or purchased at state or institutional expense. In 1990, NSF supported 59.0 percent of UNOLS's operational ship days; ONR's contribution was 15.5 percent; other federal agencies

TABLE 4-5 UNOLS Fleet

Name of Ship	Length (feet)	Built/Refit (year)	Total Ship Days (1990)
Knorr	279	1970/1991	N/A
Melville	279	1969/1992	N/A
Thompson	274	1991	N/A
Ewing	239	1983/1990	201
Vickers	220	1973/1989	N/A
Moana Wave	210	1973/1984	275
Atlantis II	210	1963	283
Wecoma	177	1975	157
Endeavor	177	1976	221
Oceanus	177	1975	239
Seward Johnson	176	1984	176
Gyre	174	1973/1980	216
New Horizon	170	1978	233
Columbus Iselin	170	1972	279
Edwin Link	168	1982/1988	107
Point Sur	135	1981	177
Cape Hatteras	135	1981	175
Alpha Helix	133	1966	109
R.G. Sproul	125	1981/1985	119
Cape Henlopen	120	1976	59
Pelican	105	1985	121
Laurentian	80	1974	148
Longhorn	80	1971/1986	53
Blue Fin	72	1972/1975	71
C.A. Barnes	65	1966/1984	154
Calanus	64	1971	93
Total ship days (1990)			4,066
Total days for ships >150 feet in length			2,680
AGOR-24	274 (planned)	?	N/A
AGOR-25	274 (planned)	?	N/A
FLIP	355	1962	65
DSRV *Alvin*		1964	241

NOTE: N/A = Not applicable; AGOR = Auxiliary General Oceanographic Research; DSRV = Deep Submergence Research Vehicle

contributed 8.6 percent; state municipalities, 10.0 percent; and foreign and private users, 6.9 percent (UNOLS, 1991). NSF's share of total funding of sea days has increased over time (Figure 4-16). The average age of the UNOLS fleet is 16.5 years (Figure 4-17). For fiscal year 1992, the total ship operations budget was about $50 million, with a larger ship costing about $15,000 per day to

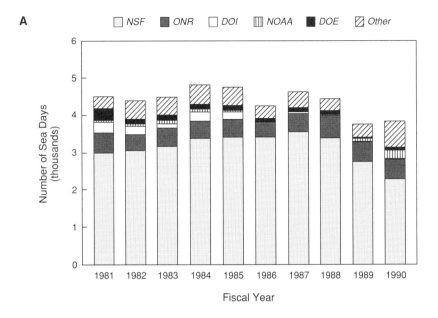

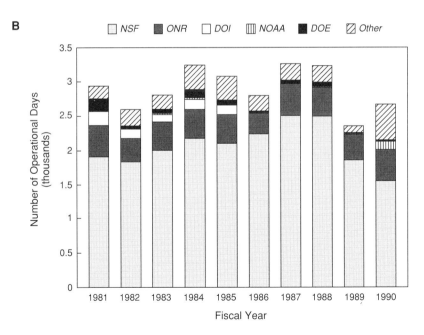

FIGURE 4-16 (A) UNOLS ship day funding by agency (all ships). From UNOLS, 1991. (B) UNOLS ship day funding by agency (ships >150 feet long). From UNOLS, 1991.

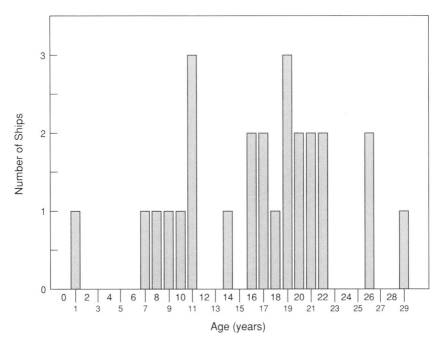

FIGURE 4-17   Age of UNOLS ships from time originally built.

operate. In addition to the UNOLS fleet, smaller vessels used primarily for coastal research are funded principally by local sources.

Federal agencies that either operate or fund oceanographic ships include the U.S. Coast Guard (USCG), the U.S. Geological Survey (USGS) and the Minerals Management Service (MMS) of the Department of the Interior (DOI), the Environmental Protection Agency (EPA), the Department of Energy (DOE), the Naval Oceanographic Office, the Office of Naval Research, the National Oceanic and Atmospheric Administration, and the National Science Foundation. The USCG is included because its two icebreakers can support research operations in the Antarctic and Arctic. The need for and operation of federal oceanographic ships arise from the statutory mission of each agency that is manifested by approved and funded programs in the federal budget. Individual agency programs dictate the requirements for ships and ship time. The federal fleet is older, on average, than the UNOLS fleet (Figure 4-18).

Ship use by different oceanography subdisciplines during the 1980s is shown in Figure 4-19. For all ships, biological oceanography uses the most ship time. For the larger ships, marine geology

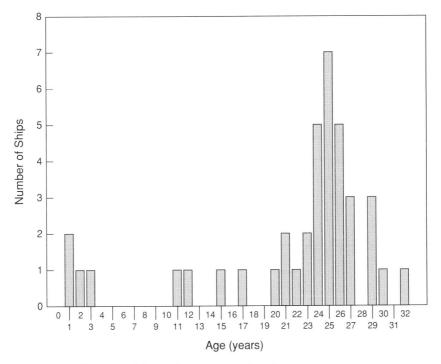

FIGURE 4-18   Age of federal oceanographic fleet.

and geophysics has the most ship days, followed by physical ocean-ography. The pattern of use, particularly for ships more than 150 feet long, is remarkably stable and probably indicates the future use of ships (Figure 4-19).

As discussed in earlier chapters, a significant development in oceanography is the increased number of large, long-term research activities planned by the academic oceanographic community. These include the Tropical Ocean-Global Atmosphere (TOGA) program, WOCE, JGOFS, and RIDGE. These major programs account for significant use of the larger ships.

Present trends suggest that research in coastal oceanography will continue to be important over the next decade because it is the primary interest of most federal mission agencies, states, and municipalities. Although some future coastal research efforts will be well served by some of the existing research vessels (*Oceanus* or *Cape* class), smaller research vessels are also needed. Specifi-cally, these new vessels should be capable of working at sea for up to 20 days at a time, at a cost of about $3,000 per day, and their

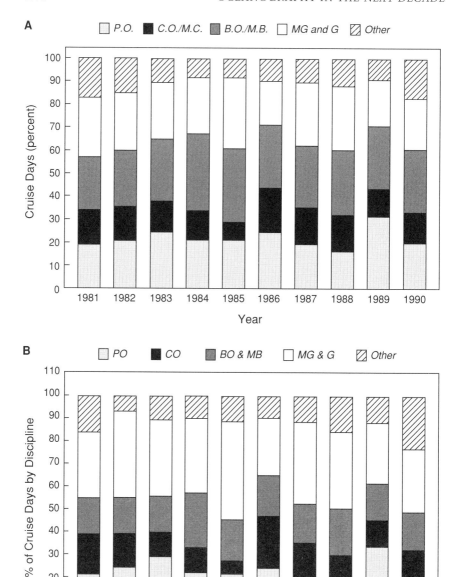

FIGURE 4-19 (A) Cruise days funded, by discipline (all UNOLS ships). (B) Cruise days funded, by discipline (UNOLS ships >150 feet long).

scheduling should be flexible. A few UNOLS vessels satisfy these criteria. Present estimates are that a vessel designed for coastal oceanography would cost $12 million to build and equip. At least one group of institutions is proceeding independently to design its own ship in this class. UNOLS is cognizant of the need for a coordinated plan to reduce any redundant effort concerning new coastal research ships.

Special Facilities

Submersibles

A broad range of submersible systems is available from the government or can be leased commercially. Since 1964 the *Alvin*, capable of operating to a depth of 4,000 meters, has given scientists a presence in the deep sea. *Alvin* is valuable to scientists who conduct research in the water column or study processes at the seawater-seafloor boundary. WHOI operates *Alvin* as a national facility, with sponsorship by an interagency agreement among NOAA, ONR, and NSF.

The Navy (Submarine Development Group One) operates the *Sea Cliff* (capable to 6,000 meters) and *Turtle* (to 3,000 meters) in support of Navy operations and research. *Sea Cliff* and *Turtle* have been used minimally by the academic community. *Sea Cliff* is the only U.S. submersible available to scientists that can operate at depths to 6,000 meters. An agreement among the Navy, NOAA, and UNOLS will improve the coordination and use of Navy deep submergence assets for academic research. Harbor Branch Oceanographic Institution owns and operates two Johnson Sea Link (1,000 meters) submersibles, which have been used intensively by academic researchers, government, and industry.

Unmanned, tethered, remotely operated vehicles (ROVs), which for some time performed ocean engineering tasks largely for the offshore oil industry, appear to be gaining acceptance and use by ocean researchers. Some ROVs are less expensive than manned submersibles and allow long submerged endurance times, making them attractive tools for some tasks vis-à-vis manned submersibles.

Floating Instrument Platform

The Floating Instrument Platform (*FLIP*), operated by SIO, fills scientific needs for a stable platform in rolling seas. It has been used for studying acoustic signals, surface and internal wave properties, and water temperature, and for collecting meteorological data. *FLIP* achieves its stability and vertical access through the water column by its extension vertically below the surface when on station. Increased efforts to improve active acoustic capabilities and all-weather operations in higher latitudes emphasize the continuing requirements for a *FLIP*-like platform. The present *FLIP* cannot support the more demanding projects expected in the near future. Research users have requested a second-generation *FLIP* that would allow for deployment of new larger, multidimensional acoustic arrays, ROVs, and other equipment under development.

JOIDES Resolution

The *JOIDES Resolution* is a specially equipped drilling vessel that has laboratory facilities for studying core samples and the capability for making downhole measurements (logging). NSF has contracted with JOI, which has in turn contracted with Texas A&M University to serve as science operator and with Lamont-Doherty Geological Observatory to provide logging and other services. The science operator is responsible for the operation of the drill ship, cruise staffing, logistics, engineering, shipboard laboratories, archiving of core samples and data, and publications. The Ocean Drilling Program is in the category of large science projects that require the application of expensive state-of-the-art technology for the advancement of the science.

Satellites: Ocean-Related Remote Sensing

In the early 1980s, NASA asked JOI to prepare a report on satellite oceanography. The report—*Oceanography from Space 1985–1995* (JOI, 1985), prepared by a committee of oceanographers expert in the field, recommended a series of ocean-related remote sensing missions that are scheduled for the 1990s (Table 4-6). It is increasingly clear that understanding the ocean is central to global change research and that the National Aeronautics and Space Administration (NASA) and analogous space agencies around the world should be major participants in the development of ocean remote sensing. Although several ocean-related missions are scheduled in the early 1990s, plans for the late 1990s and beyond are still tentative. Because of the long lead time from the concept of a satellite sensor until it is launched, efforts are needed now to ensure the development of relevant missions for the early twenty-first century to avoid gaps in time series of important measure-

TABLE 4-6 Status of Major, Pre-EOS Ocean Spacecraft and Instruments (as of fall, 1991)[a]

Satellite	Sponsor	Instruments	Launch Date
DMSP Series	USAF NASA[b]	MR	June 1987
Polar Series	NOAA NASA[b]	IR	Ongoing
ESA ERS-1	ESA NASA[b]	ALT, SCAT, SAR IR	July 1991
NASDA ERS-1	NASDA NASA[b]	SAR	February 1992
TOPEX/Poseidon	NASA CNES	ALT	July 1992
SeaWiFS	OSC NASA[b]	CS	August 1993
ESA ERS-2	ESA	ALT, SCAT, SAR IR	1994+
RADARSAT	CANADA NASA[b]	SAR	Late 1994
ADEOS	NASDA NASA[b]	SCAT, CS	1995

NOTE: ALT = radar altimeter; CNES = French space agency; CS = color scanner; ESA = European Space Agency; IR = infrared radiometer; MR = microwave radiometer; NASDA = Japanese space agency; OSC = Orbital Sciences Corporation; SAR = synthetic aperture radar; SCAT = scatterometer; SeaWiFS = Sea-viewing Wide Field Sensor.

[a]EOS = Earth Observing System.
[b]Provides data or other services to U.S. research users.

ments and deterioration of U.S. capabilities. The advancement of ocean science depends on both general Earth-observing and ocean-specific missions. The continuance and strengthening of partnerships between NASA and other agencies and with industry in the United States and abroad are key to the success of ocean-related missions.

Satellite observations contribute to studies of sea surface waves, wind speed and direction, gas fluxes, atmospheric water vapor concentrations and rainfall, sea surface temperature, ocean color, sea ice distributions, ocean surface topography, and gravity. The potential of satellite oceanography is almost unlimited, although its usefulness for most purposes depends on in situ calibration of the remote measurements.

Atmospheric water vapor must be measured because it is used in computing ocean surface evaporation and thermal forcing and is needed to correct altimetry data. Sea surface temperature observed by infrared sensors is the surface signature of ocean temperature changes. It is a vital parameter in estimating surface heat fluxes and evaporation, and can be used to infer some circulation features.

Remote sensing of ocean color is a key element for understanding the global ocean carbon budget. To obtain long-term continuous global ocean color measurements, the Sea-viewing Wide Field Sensor (SeaWiFS) sensor will be launched on a satellite in 1993. Future ocean color instruments should include improvements in spectral coverage and calibration. An ocean color sensor and scatterometer should be combined on a future satellite because of the close connection between wind stress and productivity. With future sensors, data from more wavelengths may be collected. This should allow estimation of various colored dissolved organic materials and, perhaps, separation of phytoplankton pigment groups. Sun-stimulated fluorescence at 683 nanometers (Chamberlin et al., 1990) may be a good indicator of the photosynthetic state of the phytoplankton and thus be useful in improving primary productivity models.

Passive microwave sensors measure concentrations of open water versus sea ice and may, in the future, be able to estimate the emitting temperature of the upper layer of the ice, which is related to the surface heat balance. The large-scale shape of the ocean surface (the geoid) is primarily related to Earth's gravity field because the ocean surface tends to form a level surface perpendicular to the force of gravity at any given location. Deviations from this level surface are caused primarily by ocean currents. Ocean currents can be studied by a combination of altimeter measurements of the ocean surface height and gravity measurements of the geoid.

Precise satellite geodetic measurements, providing information on crustal deformation, continental drift, and plate tectonics, Earth and ocean tides, and changes in Earth's geopotential, have been carried out since 1976 in a joint project between the United States and Italy with the Laser Geodynamics Satellite.

A mission to determine Earth's gravity field is still needed. No gravity mission is firmly in any space agency's plans, but design studies are being conducted. Of particular interest are the joint U.S./ESA (European Space Agency) plans for a gravity mission called Applications and Research Involving Space Technologies Observing the Earth's Field from Low Earth Orbiting Satellites (ARISTOTELES). The ARISTOTELES spacecraft would include both a gravity gradiometer for highly accurate gravity measurements and a magnetometer for geomagnetic studies. It is important that the geomagnetic mission begin before 1998 to avoid the next sunspot maximum, which would hamper the low-altitude initial portion of the satellite's mission.

It is clear from Table 4-6 that many objectives of the original *Space, A Research Strategy for the Decade 1985–1995* (JOI, 1985) report are being met. Yet successful completion of many missions requires more than just NASA support; new partnerships are needed. Healthy relationships between U.S. and non-U.S. space agencies and with private industry are also needed. Some of these relationships appear to be working well, for example, in Earth Resources Satellite-1 data sharing through the Alaska synthetic aperture radar facility and in the joint design of TOPEX/Poseidon with the French. Future partnerships, such as those in ocean color with the Orbital Sciences Corporation's SeaWiFS, are yet to be tested. It is clear that developing and maintaining these partnerships require strong leadership at NASA headquarters, so that U.S. participation in the process from sensor design to data analysis is guaranteed. The oceanographic community must not find itself wholly dependent on international agreements and data from non-U.S. sensors and missions during the late 1990s and beyond.

There is a need for continuing research in the development of mathematical techniques to correct satellite data for the effects of clouds, water vapor, and other atmospheric aerosols, to relate satellite measurements to observations at the ocean surface, and to relate the surface signal to processes occurring at depth. If calibration errors in the satellite data time series can be avoided, it will be possible to create a time series that is long enough to investigate low-frequency phenomena in the record of upper ocean temperatures and other variables.

Numerical Ocean Modeling

Numerical ocean modeling has reached a degree of sophistication whereby it can affect the study of present ocean circulation and the prediction of future climate. Relatively realistic multidecadal simulations of the North Atlantic, the Southern Ocean, and the world ocean have recently been carried out. The results of these experiments are being analyzed by numerous groups to aid in understanding the ocean general circulation (e.g., Boning et al., 1991; Semtner and Chervin, 1992). Data collected by comprehensive field programs such as WOCE and TOGA can be interpreted better through the use of realistic models, and field data provide essential tests for the models. WOCE is sponsoring a community modeling effort whereby different models of global circulation are compared. Overall scientific progress is maximized by the interaction of models and observations.

Future progress in modeling will involve new techniques and significantly faster computers to conduct simulations with more realistic hydrodynamics, improved resolution of eddies, longer time integration, and more testing of methods of handling subgrid-scale variables.

Technological advances will probably enhance ocean modeling more than changes in methodology. Computers are expected to attain speeds in excess of one trillion floating-point operations per second (a teraflop) before the year 2000. This thousandfold improvement over computers of 1990 will allow major improvements in simulation capability, such that realistic global models might be achieved. Their maximal use will require the development of highly parallel algorithms. Because most ocean models are formulated in terms of local space-time processes, they should be easily implemented on massively parallel computers.

The computer and communications requirements for archiving, analyzing, and visualizing the output of eddy-resolving basin- to global-scale models are vast. Ongoing federal programs in high-performance computing should help to develop some of the necessary resources. Ocean modeling was highlighted as one application of high-performance computing in the interagency Federal Coordinating Council for Science, Engineering, and Technology supplement to the president's budget for fiscal year 1993 (FCCSET, 1992). Also, large observational programs are critical because basin- to global-scale, long-term ocean data sets are required to initiate and validate ocean models.

FISCAL RESOURCES

Information on oceanographic research funding in the United States for the 11 fiscal years from 1982 to 1992 is compiled here.

NSF and ONR provide the majority of federal support for university-based basic oceanographic research. In addition, several federal mission agencies (i.e., NOAA, NASA, USGS, MMS, DOE, and EPA) support ocean science research both within their agencies and through extramural funding to the academic research community.

Federal Funding of Ocean Science

This section describes federal support of ocean science; it does not include funding by states and the private sector. For most of the mission agencies, no distinction is made between basic research conducted in a federal laboratory and that supported at universities, but for NASA, university science support is separated from total science support.

Uniform budget information for all these agencies is difficult to obtain because some agencies reorganized during fiscal years 1982–1992, and ocean and nonocean research budgets are sometimes combined into one budget category. Yearly funding is presented by agency in both current dollars (Table 4-7) and constant 1982 dollars (Table 4-8). The funding data were substantiated by the agencies for accuracy within ±5 percent. The inflation adjustment to constant dollars is based on the gross national product (GNP) index for the years 1982–1992. The GNP indices used for 1990–1992 are estimates.

The distribution of fiscal year 1992 support for basic research is shown in Figure 4-20. NSF was the largest supporter of basic oceanographic research in the United States (34.5 percent) and, along with ONR (20.4 percent) and NOAA (16.1 percent), provided more than 70 percent of the reported support in fiscal year 1992. NOAA's ocean science research programs (including Sea Grant) were funded at about the same level as the ONR program, and other federal agencies, including USGS, EPA, NASA, and MMS, have significant programs in ocean-related research. Thus to obtain a comprehensive picture of funding trends, contributions from these other federal agencies must be included.

National Science Foundation

Since the 1960s, NSF has been the principal supporter of academic oceanographers in the United States. Figure 4-21 shows the growth of the overall NSF budget and the ocean science component for fiscal years 1982–1992 in both current and constant

TABLE 4-7 Ocean Science Federal Agency Budget History: Current Dollars (millions)

Agency	Fiscal Years										
	1982	1983	1984	1985	1986	1987	1988	1989	1990	1991	1992
ONR total SE 31 and 32[a]	71.0	65.9	65.8	70.2	64.2	79.2	87.4	89.3	88.1	105.0	106.0
SE 31 ocean science	35.5	37.2	39.4	39.3	41.7	50.9	56.7	52.3	52.0	60.7	58.6
SE 32 ocean geophysics	27.5	28.1	27.7	26.2	29.2	29.2	32.1	36.1	36.1	36.2	40.9
SE 33-03 marine meteorology	N/A	N/A	N/A	N/A	N/A	N/A	N/A	N/A	N/A	N/A	6.5
Total NSF (overall)	995.6	1,093.5	1,322.6	1,501.6	1,458.3	1,622.9	1,717.0	1,923.2	2,078.8	2,366.0	2,573.0
NSF (total ocean science)[a]	95.0	102.5	114.3	121.2	119.4	133.7	135.0	145.9	147.4	164.7	178.8
OSRS	46.4	49.9	55.1	58.3	56.9	66.5	67.2	70.9	72.9	82.0	90.8
B.O.	10.7	11.8	12.9	13.9	13.3	14.4	14.8	17.1	17.3	20.3	23.1
C.O.	10.1	10.8	12.0	12.4	11.9	13.4	13.7	14.5	14.9	16.1	17.4
MG and G	12.0	12.6	14.6	15.2	14.6	16.2	16.2	16.0	16.0	17.4	19.2
P.O.	13.7	14.7	15.6	16.8	17.1	22.5	22.8	23.3	24.7	28.3	31.1
DSDP/ODP (total)	20.5	21.0	26.3	27.7	28.8	30.0	30.6	31.4	32.0	35.0	36.4
OCF	28.1	31.6	32.9	35.2	33.7	37.2	37.2	43.6	42.5	47.7	51.6
NOAA (total ocean science)[a]	117.9	133.7	91.5	83.7	86.5	75.5	73.3	80.8	81.7	84.5	83.6
Sea Grant	22.5	22.8	23.5	25.0	25.7	25.9	25.4	24.8	27.2	25.3	31.8
Global Change	—	—	—	—	—	—	—	5.4	9.5	21.2	21.2
Coastal Ocean Program	—	—	—	—	—	—	—	—	6.4	10.8	11.5
DOE (total ocean science)[a]	14.6	7.1	9.3	9.4	7.7	7.3	6.8	8.9	10.1	10.3	12.2
Oceans Research	9.9	5.0	6.6	6.7	5.0	5.4	5.1	5.8	5.8	4.5	5.5
Global Change	4.7	2.1	2.7	2.7	2.7	1.9	1.7	3.2	4.3	5.8	6.7

156

USGS[a]	21.9	13.0	18.6	21.5	25.3	26.0	28.5	29.5	32.4	37.7	36.7
MMS[a]	27.1	30.2	25.3	23.8	19.7	18.7	19.1	17.0	17.1	25.1	15.0
EPA[a]	NA	NA	NA	NA	NA	NA	NA	16.2	21.1	35.5	49.2
NASA (total ocean science)[a,b]	16.6	17.5	18.7	20.5	22.2	23.2	23.7	25.3	26.7	31.1	36.5
(including satellites)	17.4	19.1	20.6	33.7	38.0	61.0	115.6	110.5	137.7	133.8	139.4
Research and analysis[b]	16.2	17.0	18.2	19.7	20.6	20.8	21.0	22.3	22.4	25.3	26.4
University science	3.3	4.0	3.8	5.4	4.0	5.5	7.1	8.0	11.5	12.3	12.6
Flight projects total	1.2	2.1	2.4	14.0	17.4	40.2	94.6	88.2	115.3	108.5	113.0
Total nonscience	0.8	1.6	1.9	13.2	15.8	37.8	91.9	85.2	111.0	102.7	102.9
Total science[b]	0.4	0.5	0.5	0.8	1.6	2.4	2.7	3.0	4.3	5.8	10.1
Flight projects											
TOPEX/POSEIDON	1.2	2.1	2.4	3.2	4.7	9.0	68.8	76.9	96.5	78.7	62.7
(science funds)	0.4	0.5	0.5	0.6	0.6	0.6	0.8	1.0	1.8	2.0	5.3
NSCAT	—	—	—	10.3	11.7	26.2	18.3	8.0	11.4	20.6	28.3
(science funds)	—	—	—	0.1	0.5	1.2	1.2	1.2	1.2	1.3	1.6
ASF and NSIDC	—	—	—	0.5	1.0	5.0	7.4	3.1	2.9	3.2	4.5
(science funds)	—	—	—	0.1	0.5	0.6	0.7	0.8	1.3	2.5	3.2
SeaWiFS	—	—	—	—	—	—	0.1	0.2	4.5	6.0	17.5
(science funds)	—	—	—	—	—	—	0	0	0	0	0
Total federal ocean science[a]	364.0	369.9	343.5	350.3	345.0	363.6	373.8	412.9	424.6	493.9	518.0

NOTE: All 1992 values are estimates; NA = not available.

[a]Individual values are summed to obtain the total federal ocean science figure.
[b]Individual values are summed to obtain the NASA total ocean science figure.

TABLE 4-8 Ocean Science Federal Agency Budget History: Constant 1982 Dollars (millions)

Agency	Fiscal Years										
	1982	1983	1984	1985	1986	1987	1988	1989	1990	1991	1992
ONR total SE 31 and 32[a]	71.0	63.3	60.8	62.7	55.9	66.6	70.5	69.0	65.1	73.3	71.3
SE 31 ocean science	35.5	35.7	36.4	35.1	36.3	42.8	45.8	40.4	38.4	42.4	39.4
SE 32 ocean geophysics	27.5	27.0	25.6	23.4	25.4	24.6	25.9	27.9	26.7	25.3	27.5
SE 33-03 marine meteorology	NA	NA	NA	NA	NA	NA	NA	NA	NA	NA	4.4
Total NSF (overall)	995.6	1,050.4	1,221.3	1,341.9	1,269.2	1,364.9	1,385.8	1,485.1	1,536.4	1,651.1	1,730.3
NSF (total ocean science)[a]	95.0	98.5	105.5	108.3	103.9	112.4	109.0	112.7	108.9	114.9	120.2
OSRS	46.4	47.9	50.9	52.1	49.5	55.9	54.2	54.7	53.9	57.2	61.1
B.O.	10.7	11.3	11.9	12.4	11.6	12.1	11.9	13.2	12.8	14.2	15.5
C.O.	10.1	10.4	11.1	11.1	10.4	11.3	11.0	11.2	11.0	11.2	11.7
MG and G	12.0	12.1	13.5	13.6	12.7	13.6	13.1	12.4	11.8	12.1	12.9
P.O.	13.6	14.1	14.4	15.0	14.9	18.9	18.4	18.0	18.3	19.7	20.9
DSDP/CDP (total)	20.5	20.2	24.3	24.8	25.1	25.2	24.7	24.2	23.7	24.4	24.5
OCF	28.1	30.4	30.4	31.5	29.3	31.3	30.0	33.7	31.4	33.3	34.7
NOAA (total ocean science)[a]	117.9	128.4	84.5	74.8	75.3	63.5	59.2	62.4	60.4	59.0	56.2
Sea Grant	22.5	21.9	21.7	22.3	22.4	21.8	20.5	19.2	20.1	17.7	21.4
Global Change	—	—	—	—	—	—	—	4.2	7.0	14.8	14.3
Coastal Ocean Program	—	—	—	—	—	—	—	—	4.7	7.5	7.7
DOE (total ocean science)[a]	14.6	6.8	8.6	8.4	6.7	6.1	5.5	6.9	7.5	7.2	8.2
Oceans Research	9.9	4.8	6.1	6.0	4.4	4.5	4.1	4.5	4.3	3.1	3.7
Global Change	4.7	2.0	2.5	2.4	2.3	1.6	1.4	2.5	3.2	4.0	4.5

USGS[a]	21.9	12.5	17.2	19.2	22.0	21.9	23.0	22.8	23.9	26.3	24.7
MMS[a]	27.0	29.0	23.4	21.3	17.1	15.7	15.4	13.1	12.6	17.5	10.1
EPA[a]	NA	NA	NA	NA	NA	NA	NA	12.5	15.6	24.8	33.1
NASA (total ocean science)[ab]	16.6	16.8	17.3	18.3	19.3	19.5	19.1	19.5	19.7	21.7	24.5
(including satellites)	17.4	18.3	19.0	30.1	33.1	51.3	93.3	85.3	101.8	93.4	93.7
Research and analysis[b]	16.2	16.3	16.8	17.6	17.9	17.5	16.9	17.2	16.6	17.7	17.8
University science	3.3	3.8	3.5	4.8	3.5	4.6	5.7	6.2	8.5	8.6	8.5
Flight projects total	1.2	2.0	2.2	12.5	15.1	33.8	76.4	68.1	85.2	75.7	76.0
Total nonscience	0.8	1.5	1.7	11.8	13.7	31.8	74.2	65.8	82.0	71.7	69.2
Total science[b]	0.4	0.5	0.5	0.7	1.4	2.0	2.2	2.3	3.2	4.0	6.8
Flight projects											
TOPEX/POSEIDON	1.2	2.0	2.2	2.9	4.1	7.6	55.5	59.4	71.3	54.9	42.2
(science funds)	0.4	0.5	0.5	0.5	0.5	0.5	0.6	0.8	1.3	1.4	3.6
NSCAT	—	—	—	9.2	10.2	22.0	14.8	6.2	8.4	14.4	19.0
(science funds)	—	—	—	0.1	0.4	1.0	1.0	0.9	0.9	0.9	1.1
ASF and NSIDC	—	—	—	0.4	0.9	4.2	6.0	2.4	2.1	2.2	3.0
(science funds)	—	—	—	0.1	0.4	0.5	0.6	0.6	1.0	1.7	2.2
SeaWiFS	—	—	—	—	—	—	0.1	0.2	3.3	4.2	11.8
(science funds)	—	—	—	—	—	—	0.0	0.0	0.0	0.0	0.0
Total federal ocean science[a]	364.0	355.3	317.3	313.0	300.2	305.7	301.7	318.9	313.7	344.7	348.3

NOTE: All 1992 values are estimates; NA = not available.

[a]Individual values are summed to obtain the total federal ocean science figure.
[b]Individual values are summed to obtain the NASA total ocean science figure.

159

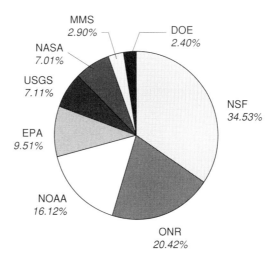

FIGURE 4-20 Distribution of federal support for ocean science in Fiscal Year 1992.

1982 dollars. The NSF budget grew at an annual rate of 14.4 percent during this time. More than half of this increase can be attributed to inflation; in constant 1982 dollars, the total NSF budget increased at an annual rate of 6.7 percent. This impressive record indicates continuing support in both the administration and the Congress for basic scientific research.

Ocean Science   The budget of NSF's Ocean Sciences Division (OCE) has not increased as rapidly as the overall NSF budget over this same period (Figure 4-21). In constant 1982 dollar terms, the OCE budget grew 2.4 percent annually between fiscal years 1982 and 1992, a constant dollar growth rate about one-third that of the overall NSF budget. Of the OCE growth, in constant 1982 dollar terms, 58 percent can be attributed to growth specifically in Ocean Science Research Support (OSRS). The Ocean Drilling Program (ODP) accounts for 16 percent of the constant 1982 dollar growth and Oceanographic Centers and Facilities (OCF) for 26 percent. It is encouraging to note that the 5.5 percent increase in the OCE budget from fiscal years 1990 to 1991 (in constant 1982 dollars) and the 4.6 percent budget increase from 1991 to 1992 may signal significant real growth in the OCE budget in the 1990s.

Funding increases have not been uniform across the oceanographic disciplines in OCE (Figure 4-22). The physical oceanography budget increased more than the other three disciplines, ac-

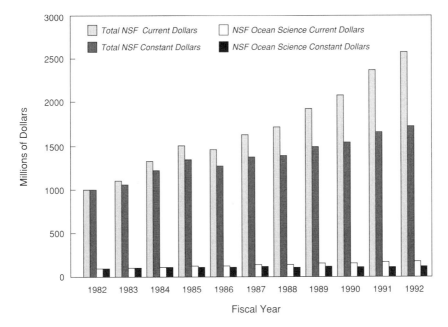

FIGURE 4-21 Budget history of the National Science Foundation and the ocean science component in current and constant 1982 dollars for Fiscal Years 1982–1992.

counting for 53 percent (in inflation-adjusted dollars) of the entire OSRS growth between fiscal years 1982 and 1992. Biological oceanography accounted for 33 percent of the OSRS growth. In contrast, increases in the chemical oceanography and marine geology and geophysics budgets accounted for much smaller percentages of the OSRS growth, 11 and 6 percent, respectively. However, this relatively slow growth in core program support for MG and G has been offset by a $5 million to $6 million budget per year for drilling-related science that began when ODP was established in the mid-1980s.

Thus at NSF, 1982–1992 was characterized by slow growth in research support for ocean sciences. Further, the percentage growth occurred mostly in OSRS and can be attributed primarily to increased support in physical oceanography and, in fiscal years 1991 and 1992, biological oceanography as well.

Other Basic Sciences Overall, NSF support for most fields of basic scientific research grew relatively slowly from fiscal years 1982 to 1992. The three directorates that fund most of NSF's

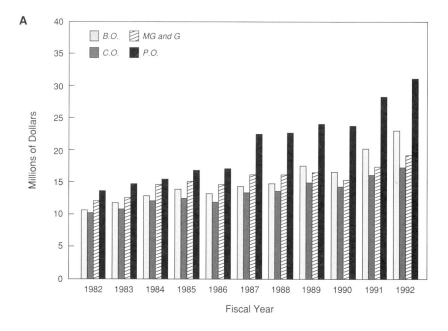

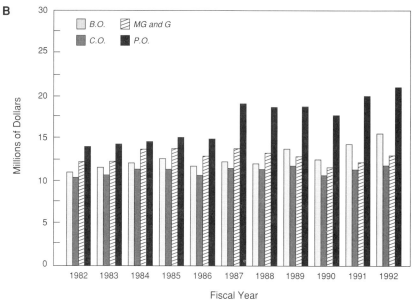

FIGURE 4-22 Budget history of the National Science Foundation's ocean science disciplines in current dollars (A) and in constant 1982 dollars (B) for Fiscal Years 1982–1992.

basic scientific research (and comprise more than one-half its to-tal budget)—Biological, Behavioral and Social Sciences, Mathematics and Physical Sciences, and Geosciences—had budget growth rates substantially lower than the overall NSF budget. NSF director-ates responsible for technology, computing, engineering, and edu-cation accounted for most of the percentage growth in the overall NSF budget.

Office of Naval Research

The Department of the Navy, primarily through the ONR, has been a major supporter of basic oceanographic research in the United States. ONR funding has changed little in constant dol-lars since fiscal year 1982 (Figure 4-23). Funding by ONR's oceano-graphic disciplines, which differ from NSF's, are also relatively constant (Figure 4-23).

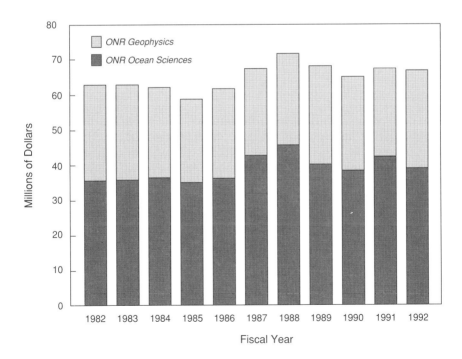

FIGURE 4-23 Office of Naval Research funding for ocean science in constant 1982 dollars for Fiscal Years 1982–1992.

Office of the Oceanographer of the Navy

The Office of the Oceanographer of the Navy was the program sponsor for the following new construction of Navy-owned ships assigned to academic institutions between fiscal years 1982 and 1992:

$33 million	AGOR-23 (R/V *Thompson*)	New construction
$47 million	R/V *Knorr*, R/V *Melville*	Refitting
$41 million	AGOR-24	New construction

Other Navy Support

Other Navy support for ocean science comes from the Office of Naval Technology (ONT) and the Naval Research Laboratory (NRL). ONT provided $43.7 million in fiscal year 1992 for science, but no breakdown for ocean science is available. Further, no budget figures are available prior to fiscal year 1992. NRL provided $3.2 million in fiscal year 1992 for ocean science; here too, no prior budget figures are available yet.

National Oceanic and Atmospheric Administration

NOAA's research budget includes mapping, charting, geodesy activities, ocean and coastal management, climate research, and fisheries management (Figure 4-24). NOAA research is carried out at major federal laboratories, such as the Atlantic Oceanographic and Meteorological Laboratories and the Pacific Marine Environmental Laboratories, as well as through cooperative agreements with universities and the National Sea Grant College, Climate and Global Change, and Coastal Ocean programs.

Sea Grant, NOAA's major extramural funding program for university-based scientists, provided approximately $25.3 million in fiscal year 1991 for ocean science research (Figure 4-25). The Climate and Global Change Program began in fiscal year 1989 and provides some support for academic scientists (Figure 4-25). The Coastal Ocean Program (COP) began in fiscal year 1990. Approximately 50 percent of its $11.5 million budget for fiscal year 1992 is used to support academic research in coastal ocean science (Figure 4-25). Although it is a young program, COP indicates a possible trend of increasing academic research support (164 percent between fiscal years 1990 and 1992 in constant 1982 dollars). If its budget continues to increase and congressional support continues, COP may emerge as a significant extramural funding pro-

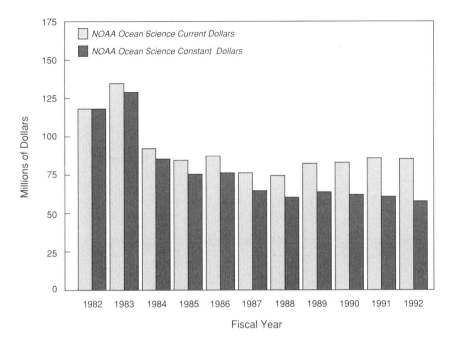

FIGURE 4-24 National Oceanic and Atmospheric Administration (NOAA) funding for ocean science in current and constant 1982 dollars for Fiscal Years 1982–1992.

gram in the 1990s. Funding information for these three NOAA programs is included in Tables 4-7 and 4-8.

Department of Energy

For many years, the Department of Energy has supported a marine research program in areas such as subseabed waste disposal, carbon dioxide-related research, and coastal oceanography (Figures 4-26 and 4-27). In fiscal year 1982, the marine research program was budgeted at $22.9 million, with some of the work contracted to university-based marine scientists. Between fiscal years 1982 and 1987, the budget was reduced nearly 75 percent in constant 1982 dollar terms. Programs in subseabed waste disposal and strategic petroleum were eliminated, and funding for coastal oceanography and carbon dioxide research was reduced. With DOE involvement in the U.S. Global Change Program, funding for carbon dioxide-related research has rebounded. Since fis-

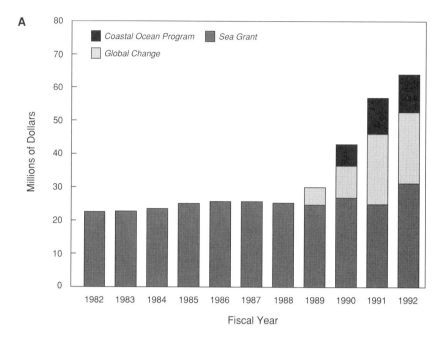

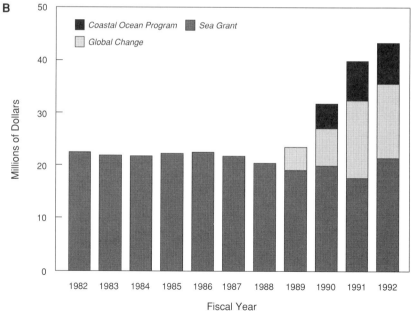

FIGURE 4-25 Budget history of the National Oceanic and Atmospheric
Administration's Sea Grant, Coastal Ocean Program, and Global Change
ocean science components in current dollars (A) and in constant 1982
dollars (B) for Fiscal Years 1982–1992.

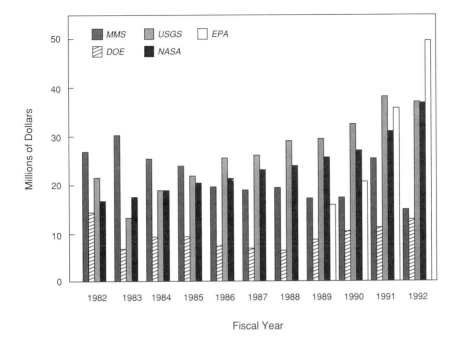

FIGURE 4-26 Budget history of ocean science research programs in several major federal mission agencies in current dollars for Fiscal Years 1982–1992.

cal year 1987, DOE funding for ocean-related research increased 6.6 percent annually in constant 1982 dollar terms, but it is still significantly (63 percent) below the level of fiscal year 1982 support in constant 1982 dollars.

U.S. Geological Survey

USGS supports marine geological and geophysical research. During the past decade, it has emphasized mapping and assessing the geological resources of the U.S. Exclusive Economic Zone. USGS ocean science funding—which includes two major components, Offshore Geologic Framework and Coastal Geology—decreased 32 percent in constant 1982 dollars from fiscal year 1982 to 1983 (Figures 4-26 and 4-27). This reduction is due in part to the formation of a new bureau MMS, which was separated from the Conservation Division unit in the Department of the Interior in fiscal year 1982. Since fiscal year 1983, the USGS marine programs budget has grown 65.8 percent in constant 1982 dollar terms, a 6.6 percent annual average increase.

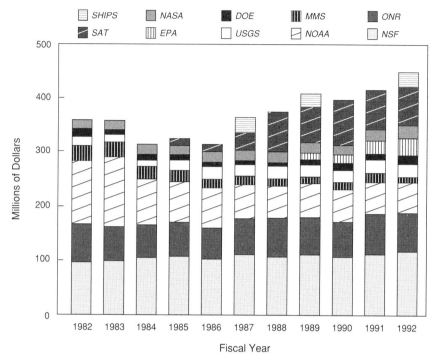

FIGURE 4-27 Total federal support for ocean science including NASA satellites and Navy ships assigned to academic institutions, in constant 1982 dollars for Fiscal Years 1982–1992.

Minerals Management Service

MMS's Environmental Studies Program supports studies in physical oceanography, offshore geology, and marine pollution. Some studies are contracted with university-based researchers, and others are conducted by private industry or federal agencies (e.g., USGS). In general, the MMS ocean science budget has decreased continuously from fiscal year 1982 to fiscal year 1992, for a 63 percent overall decrease in constant 1982 dollars in these 11 years (Figures 4-26 and 4-27).

Environmental Protection Agency

EPA has a rapidly growing marine research program. Reliable figures are not available prior to fiscal year 1989, but between fiscal years 1989 and 1992, the EPA marine program budget increased 165 percent in constant 1982 dollars (an average annual increase of 41 percent), the largest percent increase in any federal agency (Figures 4-26 and 4-27).

National Aeronautics and Space Administration

Satellites are increasingly important in modern oceanographic research. NASA provides funding for construction, operation, and related research for ocean satellite missions such as TOPEX/Poseidon, instruments such as SeaWiFS and the NASA Scatterometer, and data collection and analysis from other satellites such as ESA's Earth Resources Satellite-1 (see "Physical Resources"). It is difficult from NASA's budget presentation to identify specific ocean-related funding after fiscal year 1989, except for individual satellites. Expenditures for fiscal years 1982–1992 are shown in Tables 4-7 and 4-8 in two categories, Research and Analysis and Flight Programs. Funding of university-based researchers has nearly quadrupled in current dollars, from $3.3 million in 1982 to $12.6 million in 1992. NASA's ocean-related funding has grown, particularly for the development of new satellite sensors. Growth of NASA's budget in Earth observations is expected to be substantial as the Mission to Planet Earth begins and the Earth Observing System satellites are developed.

Discussion

Overall, federal funding of oceanographic research in the 1980s was relatively constant. Figure 4-27 shows that total federal spending on oceanographic research grew 5.1 percent from fiscal year 1982 to fiscal year 1992 (in constant 1982 dollars), an increase of about 0.6 percent annually.

Although this report focuses on funding trends in the ocean sciences, funds for individual oceanographic investigators are influenced by the rapid growth in the number of academic oceanographers and a significant increase in the costs of ocean science. Throughout the period of slow growth in federal spending on the ocean sciences in the 1980s, the number of scientists competing for funds continued to grow. According to the OSB survey (see "Human Resources"), the number of Ph.D.-level academic ocean scientists increased about 70 percent from 1980 to 1990. WHOI data indicate that the number of proposals per staff member increased from 2.8 in 1975 to 4.8 in 1991. This finding seems to confirm a general impression among research oceanographers that they now spend more time writing proposals than in the past.

The costs of the latest equipment (e.g., ships, satellites, and laboratory instrumentation) used in oceanography today are rising much faster than the rate of inflation. This trend, seen in many scientific fields, is what D. Allan Bromley, the President's Science Advisor, calls the sophistication factor. For example, all major oceanographic research vessels in the 1970s were equipped with wide-beam echo sounders to measure the water depth beneath the ship. These simple systems cost a few thousand dollars to install and were inexpensive to operate. In the 1980s, the first multiple narrow-beam echo sounders were introduced. These systems produced more accurate seafloor maps up to 16 times faster than the older echo sounders, but they cost nearly $1 million per ship to install and are much more costly to operate and maintain. In the early 1990s, the second-generation multibeam swath mapping systems were introduced. They are up to 10 times faster than the first multibeam systems but cost nearly 2.5 times as much. This example is not atypical; each oceanography discipline could cite similar examples. As our capability to do oceanographic research has increased over the past 20 years, the associated costs of acquiring, operating, and maintaining modern facilities and equipment have outpaced inflation.

References

Altabet, M.A. 1989. Particulate new nitrogen fluxes in the Sargasso Sea. Journal of Geophysical Research 94:12771-12779.

Battisti, D.S., and B.M. Hickey. 1984. Application of remote wind-forced coastal trapped wave theory to the Oregon and Washington coasts. Journal of Physical Oceanography 14:887-903.

Baumgartner, A., and E. Reichel. 1975. The World Water Balance. Amsterdam, Holland: Elsevier, 179 pp.

Berger, W.H. 1988. Global maps of ocean productivity. Pp. 429-455 in Productivity of the Ocean: Present and Past, W. H. Berger, V. S. Smetacek, and G. Wefer, eds. New York: John Wiley & Sons.

Boning, C.W., R. Doscher, and R.G. Budich. 1991. Seasonal transport variation in the western subtropical North Atlantic: Experiments with an eddy-resolving model. Journal of Physical Oceanography 21:1271-1289.

Boyle, E.A. 1990. Quaternary deepwater paleoceanography. Science 249: 863-870.

Brassell, S.C., G. Eglinton, I.T. Marlowe, U. Pflaumann, and M. Sarnthein. 1986. Molecular stratigraphy: A new tool for climatic assessment. Nature 320:129-133.

Bush, V. 1945. Science—The Endless Frontier. A report to the president on a program of postwar scientific research. Washington, D.C.: U.S. Government Printing Office, 184 pp.

Chamberlin, W.S., C.R. Booth, D.A. Kiefer, J.H. Morrow, and R.C. Murphy. 1990. Deep-Sea Research 37:951-973.

Cushing, D.H. 1982. Climate and Fisheries. New York, N.Y.: Academic Press, 373 pp.

Detrick, R.S., P. Buhl, E. Vera, J. Mutter, J. Orcutt, J. Madsen, and T. Brocher. 1987. Multichannel seismic imaging of a crustal magma chamber along the East Pacific Rise. Nature 326:35-41.

Dickson, R.R., J. Meincke, S.-A. Malmberg, and A. Jake. 1988. The great salinity anomaly. Progress in Oceanography 20:103-151.

Environmental Protection Agency Science Advisory Board. 1990. Reducing Risk. Washington, D.C.: U.S. Government Printing Office, 33 pp. (appendixes, 361 pp).

Federal Coordinating Council for Science, Engineering, and Technology. 1992. Grand Challenges: High Performance Computing and Communications. The FY 1993 U.S. Research and Development Program. A Report by the Committee on Physical, Mathematical, and Engineering Sciences. Washington, D.C.: U.S. Government Printing Office.

Geernaert, G. 1990. Bulk parameterizations for the wind stress and heat fluxes. Pp. 91-172 in Surface Waves and Fluxes: Theory, Vol. 1, G. Geernaert and W. Plant, eds. Netherlands: Kluwer Academic Publishers.

Gordon, A.L., R.F. Weiss, W.M. Smethie, Jr., and M.J. Warner. 1992. Thermocline and intermediate water communication between the South Atlantic and Indian Oceans. Journal of Geophysical Research 97:7223-7240.

Joint Oceanographic Institutions Satellite Committee. 1985. Oceanography from Space. A Research Strategy for the Decade 1985-1995. Part 1, Executive Summary; Part 2, Proposed Measurements and Missions. Washington, D.C.: JOI, Inc.

Kennett, J.P. 1977. Cenozoic evolution of Antarctic glaciation, the circumantarctic ocean and their impact on global paleoceanography. Journal of Geophysical Research 82:3843-3860.

Kullenberg, G. 1986. Long-term changes in the Baltic ecosystem. Pp. 19-32 in Variability and Management of Large Marine Ecosystems, K. Sherman and L.M. Alexander, eds. Washington, D.C.: American Association for the Advancement of Science Selected Symposium 99.

Lampitt, R.S. 1990. Directly measured rapid growth of a deep-sea barnacle. Nature 345:805-807.

Larson, R.L. 1991. Geological consequences of superplumes. Geology 19:963-966.

Lee, T. N., J. A. Yoder, and L. P. Atkinson. 1991. Gulf Stream frontal eddy influence on productivity of the southeast U.S. continental shelf. Journal of Geophysical Research 96:191-205.

Manabe, S., and R.J. Stouffer. 1988. Two stable equilibria of a coupled ocean-atmosphere model. Journal of Climate 1:841-866.

Mantoura, R.F.C., Martin, J.M., and R. Wollast, eds. 1991. Ocean margin processes in global change. Dahlem Workshop Reports; Physical,

Chemical, and Earth Sciences Research Report 9. March 18-23, 1990, Berlin.

McGowan, J.A. 1990. Climate and change in oceanic ecosystems: The value of time-series data. Trends in Ecology and Evolution 5:293-299.

Moore III, B., and B. Bolin. 1986. The oceans, carbon dioxide and global climate change. Oceanus 29:9-15.

National Research Council. 1970. Growth and Support of Oceanography in the United States. Washington, D.C.: National Academy Press, 44 pp.

National Research Council. 1971. Ocean Science Manpower Data and Their Interpretation. Washington, D.C.: National Academy Press, 19 pp.

National Research Council. 1972. Ocean Science Graduate Students. Washington, D.C.: National Academy Press, 18 pp.

National Research Council. 1981. Doctoral Scientists in Oceanography. Washington, D.C.: National Academy Press, 155 pp.

National Science Foundation. 1975. Characteristics of Doctoral Scientists and Engineers in the United States: 1973. NSF 75-312. Washington, D.C.

National Science Foundation. 1977. Characteristics of Doctoral Scientists and Engineers in the United States: 1975. NSF 77-309. Washington, D.C.

National Science Foundation. 1979. Characteristics of Doctoral Scientists and Engineers in the United States: 1977. NSF 79-306. Washington, D.C.

National Science Foundation. 1981. Characteristics of Doctoral Scientists and Engineers in the United States: 1979. NSF 80-323. Washington, D.C.

National Science Foundation. 1983. Characteristics of Doctoral Scientists and Engineers in the United States: 1981. NSF 82-332. Washington, D.C.

National Science Foundation. 1985. Characteristics of Doctoral Scientists and Engineers in the United States: 1983. NSF 85-303. Washington, D.C.

National Science Foundation. 1987. Characteristics of Doctoral Scientists and Engineers in the United States: 1985. NSF SRS-86-D3. Washington, D.C.

National Science Foundation. 1989. Characteristics of Doctoral Scientists and Engineers in the United States: 1987. NSF 88-331. Washington, D.C.

National Science Foundation. 1991. Characteristics of Doctoral Scientists and Engineers in the United States: 1989. NSF 91-317. Washington, D.C.

Nichols, F.H. 1985. Abundance fluctuations among benthic invertebrates in two Pacific estuaries. Estuaries 8:136-144.

O'Brien, J.J., ed. 1985. Advanced Physical Oceanographic Numerical Modelling. Boston: Dordrecht, 608 pp.

Office of Management and Budget. 1992. Budget of the United States Government. Fiscal Year 1993. Washington, D.C.: U.S. Government Printing Office.

Prahl, F.G., and S.G. Wakeham. 1987. Calibration of unsaturation patterns in long-chain ketone compositions for paleotemperature assessment. Nature 330:367-369.

Rona, P.A., G. Klinkhammer, T.A. Nelson, J.H. Trefry, and H. Elderfield. 1986. Black smokers, massive sulphides, and vent bacteria at the Mid-Atlantic Ridge. Nature 321:33-37.

Rothschild, B. J., and P. R. Osborn. 1988. Small-scale turbulence and plankton contact rates. Journal of Plankton Research 10(3):465-474.

Semtner, A.J., and R.M. Chervin. 1992. Ocean general circulation from a global eddy-resolving model. Journal of Geophysical Research 97: in press.

Shackleton, N.J. 1987. Oxygen isotopes, ice volumes, and sea level. Quaternary Science Reviews 6:183-190.

Sissenwine, M.P. 1986. Perturbation of a predator-controlled continental shelf ecosystem. Pp. 55-86 in Variability and Management of Large Marine Ecosystems, K. Sherman and L.M. Alexander, eds. Washington D.C.: American Association for the Advancement of Science Selected Symposium 99.

Stewart, R. H. 1985. Methods of Satellite Oceanography. Berkeley: University of California Press, 360 pp.

Stommel, H. 1961. Thermohaline convection with two stable regimes of flow. Tellus 13(2):224-230.

Tans, P.P., I.Y. Fung, and T. Takahashi. 1990. Observational constraints on the global atmospheric CO_2 budget. Science 247:1431-1438.

University-National Oceanographic Laboratory Systems. 1991. Summary Report of the UNOLS Annual Meeting. October 17, 1991, Washington, D.C.

Walsh, J.J. 1991. Importance of continental margins in the marine biogeochemical cycling of carbon and nitrogen. Nature 350:53-55.

Watson, A.J., and J.R. Ledwell, 1988. Purposefully released tracers. Philos. Trans. R. Soc. London Ser. A 325:189-200.

Weyl, P.K. 1968. The role of the oceans in climate change: A theory of the ice ages. Meteorological Monographs 8:37-62.

World Ocean Circulation Experiment Scientific Steering Group. 1986. Scientific Plan for the World Ocean Circulation Experiment. World Climate Research Programme, WCRP Publications Series, No. 6, Geneva, Switzerland: World Meteorological Organization WMO/TD- No. 122, 83 pp.

Yoder, J.A., L.P. Atkinson, T.N. Lee, H.H. Kim, and C.R. McClain. 1981. Role of Gulf Stream frontal eddies in forming phytoplankton patches on the outer southeastern shelf. Limnology and Oceanography 26:1103-1110.

Zoback, M.D., D. Moss, L. Mastin, and R.N. Anderson. 1985. Well-bore breakouts and in situ stress. Journal of Geophysical Research 90:5523-5530.

APPENDIXES

I

How the Study Was Conducted

This study is the result of several years' activity by the Ocean Studies Board (OSB) and the U.S. ocean science community. The Board convened three workshops, one on facilities (May 30-31, 1990) and two on future science directions in oceanography (March 11-12 in Irvine, California and April 22, 1991 in Washington, D.C.). Additional community input was sought through two special sessions at meetings of the American Geophysical Union. In addition, the OSB surveyed federal agencies and academic institutions on fiscal, physical, and human resources.

The OSB especially thanks scientists who were not members of the Board during the duration of this study but who contributed to or reviewed portions of this report or contributed at one of its workshops:

Alice Alldredge	Harry Bryden	Hugh Ducklow
Neil Andersen	David Christie	Ann Durbin
David Aubrey	Thomas Church	John Edmond
Arthur Baggeroer	Michael Coffin	J. Farrell
D. James Baker	Russ Davis	Rana Fine
Karl Banse	John Delaney	Fred Fisher
William Berggren	Roland deSzoeke	Jeff Fox
Michael Brown	Tom Dickey	Gary L. Geernaert
Otis Brown	Andrew Dickson	Wayne R. Geyer

Joel Goldman
Jeff Graham
George Grice
Melinda Hall
David Halpern
Eric Hartwig
James Hays
Thomas Hayward
G. Ross Heath
John Imbrie
David Johnson
Kenneth Johnson
Peter Jumars
Keith Kaulum
James Kennett
Dana R. Kester
Victor Klemas
Devendra Lal
Richard Lambert
Donald Langenberg
Charles Langmuir
Art Lerner-Lam
James Ledwell
Bonnie MacGregor
Curt Mason

Marsha McNutt
Joan Mitchell
Ralph Moberly
Christopher Mooers
J. Bradford Mooney
Jason Phipps Morgan
Walter Munk
John Mutter
Stewart Nelson
William A. Nierenberg
Peter Niiler
Charles Nittrouer
Worth Nowlin
E. Okal
Donald Olson
William Patzert
Charles Peterson
Michael Pilson
Robert Pinkel
Nicolas Pisais
Robert Presley
Joseph Prospero
Barry Raleigh
Desiraju B. Rao
Roger Revelle

Peter Rhines
Steve Riser
Paola Rizzoli
Bruce Rosendahl
Thomas Rossby
George Saunders
David Schink
Ronald Schlitz
Raymond W. Schmitt
Jerry Schubel
Bert Semtner
Thomas Shipley
Eugene A. Silva
Michael P. Sissenwine
George Somero
Derek W. Spencer
William Stubblefield
Fumiko Tajima
Taro Takahashi
Ronald Tipper
Brian E. Tucholke
Pat Walsh
Clinton Winant
Xiao-Hai Yan
James Yoder

II

Abbreviations and Acronyms

ACOS	Advisory Committee on Ocean Sciences (ACOS)
ALT	Radar altimeter
ARISTOTELES	Applications and Research Involving Space Technologies Observing the Earth's Field from Low Earth Orbiting Satellites
ARM	Atmospheric Radiation Measurements program (DOE)
BLM	Bureau of Land Management
B.O.	Biological oceanography
CHAMMP	Computer Hardware, Advanced Modeling and Model Physics program
CNES	French space agency
C.O.	Chemical oceanography
COA	Council on Ocean Affairs
COP	Coastal Ocean Program (NOAA)
CS	Color scanner
DIN	dissolved inorganic nitrogen
DOC	dissolved organic carbon
DOD	Department of Defense
DOE	Department of Energy
DOI	Department of the Interior

DOM	dissolved organic material
DON	dissolved organic nitrogen
DSDP	Deep Sea Drilling Project
EEZ	Exclusive Economic Zone
EOS	Earth Observing System
EOSDIS	EOS Data and Information System
EPA	Environmental Protection Agency
ERS-1	Earth Resources Satellite-1 (European Space Agency)
ESA	European Space Agency
FCCSET	Federal Coordinating Council for Science, Engineering, and Technology
FLIP	Floating Instrument Platform
FY	Fiscal Year
GLOBEC	Global Ocean Ecosystems Dynamics
GNP	gross national product
GOOS	global ocean observing system
IR	Infrared radiometer
JGOFS	Joint Global Ocean Flux Study
JOI	Joint Oceanographic Institutions, Inc.
JOIDES	Joint Oceanographic Institutions for Deep Earth Sampling
M.B.	Marine biology
M.C.	Marine chemistry
MG and G	Marine geology and geophysics
MIT	Massachusetts Institute of Technology
MMS	Minerals Management Service (DOI)
MR	Microwave radiometer
NADW	North Atlantic Deep Water
NASA	National Aeronautics and Space Administration
NASDA	Japanese space agency
NOAA	National Oceanic and Atmospheric Administration
NODC	National Oceanographic Data Center
NRC	National Research Council
NRL	Naval Research Laboratory
NSF	National Science Foundation
OCE	Division of Ocean Sciences (NSF)
OCF	Oceanographic Centers and Facilities (NSF)
OCS	outer continental shelf
ODP	Ocean Drilling Program
O.E.	Ocean engineering

ONR	Office of Naval Research
ONT	Office of Naval Technology
OSB	Ocean Studies Board
OSC	Orbital Sciences Corporation
OSRS	Ocean Science Research Support (NSF)
PI	principal investigator
P.O.	Physical oceanography
POM	particulate organic material
RIDGE	Ridge Inter-Disciplinary Global Experiment
ROV	remotely operated vehicle
R/V	research vessel
SAR	synthetic aperture radar
SCAT	Scatterometer
SeaWiFS	Sea-viewing Wide Field Sensor; an ocean color satellite instrument
SIO	Scripps Institution of Oceanography
TOGA	Tropical Ocean-Global Atmosphere program
TOPEX/Poseidon	Joint NASA/French Space Agency venture to measure the surface topography of the ocean with great precision
UH	University of Hawaii
UNOLS	University-National Oceanographic Laboratory System
USCG	U.S. Coast Guard
USGS	U.S. Geological Survey (DOI)
UW	University of Washington
VOS	volunteer observing ships
WHOI	Woods Hole Oceanographic Institution
WOCE	World Ocean Circulation Experiment

III

Recent Workshop and Other Reports Relevant to Discussion in "Directions in Biological Oceanography"

Atlantic Climate Change Program Science Plan. 1990. NOAA Climate and Global Change Program Special Report No. 2. University Corporation for Atmospheric Research, Boulder, Colorado, 29 pp.

Atmosphere-Ocean Exchange of Carbon Dioxide: Implications for Climate and Global Change on Seasonal-to-Century Time-Scales. 1990. NOAA Climate and Global Change Program Special Report No. 3. University Corporation for Atmospheric Research, Boulder, Colorado, 31 pp.

Brink, K.H. and others. 1990. Coastal Ocean Processes (CoOP): Results of an Interdisciplinary Workshop. Contribution No. 7584, Woods Hole Oceanographic Institution, Woods Hole, Massachusetts, 51 pp.

Coastal Ocean Margin Flux Study (COMFS). 1988. A new DOE research initiative. Marine Research Program, Office of Energy Research, Department of Energy, Washington, D.C., 25 pp.

Deep Sea Observatories. 1989. Report of a conference to assess near-term opportunities and long-range goals of deep-sea observatories (DSOs). Woods Hole Oceanographic Institution, Woods Hole, Massachusetts, 54 pp.

Dynamics of the continental margins: A report to the U.S. Department of Energy. 1990. Report NTIS-PR-360, National Technical Information Service, Springfield, Virginia, 55 pp.

Eden, H.F., and C.N.K. Mooers. 1990. Coastal Ocean Prediction Systems. Synopsis. JOI, Inc., Washington, D.C., 20 pp.

Global Ocean Ecosystems Dynamics. 1988. Report of a workshop on global ocean ecosystems dynamics. JOI, Inc., Washington, D.C., 131 pp.

Global Ocean Ecosystems Dynamics. 1991. GLOBEC Workshop on Biotechnology Applications to Field Studies of Zooplankton. Report Number 3. JOI, Inc., Washington, D.C.

Global Ocean Ecosystems Dynamics, 1991. Initial Science Plan. Report Number 1. JOI, Inc., Washington, D.C., 93 pp.

Global Ocean Ecosystems Dynamics. 1991. Theory and Modeling in GLOBEC: A first step. JOI, Inc., Washington, D.C.

Global Ocean Observing Systems Workshop Report. In press. JOI, Inc., Washington, D.C.

International Geosphere-Biosphere Programme (IGBP). 1990. Coastal Ocean Fluxes and Resources. IGBP Report Number 14. 53 pp.

Joint Global Ocean Flux Study. 1989. Report of the JGOFS Pacific Planning

Workshop, Honolulu. JGOFS Report Number 3. Scientific Committee on Oceanic Research-International Council of Scientific Unions. 68 pp.

Joint Global Ocean Flux Study Science Plan. 1990. JGOFS Report No. 5. Scientific Committee on Oceanic Research-International Council of Scientific Unions.

Joint Oceanographic Institutions, Inc. 1990. Initiatives for the accelerated transfer of biotechnology to the ocean sciences. Report of a workshop held September 16-18, 1988. Tucson, Arizona, 40 pp.

Recruitment Processes and Ecosystem Structure of the Sea. 1987. A report of a workshop. National Academy Press. Washington, D.C., 42 pp.

Sarachik, E.S., and R.H. Gammon. 1989. The role of the ocean in the NOAA program "Climate and Global Change." NOAA Climate and Global Change Program Special Report No. 1. University Corporation for Atmospheric Research, Boulder, Colorado, 49 pp.

U.S. Joint Global Ocean Flux Study Long Range Plan. 1990. U.S. JGOFS Planning Report Number 11. U.S. JGOFS Planning Office, Woods Hole Oceanographic Institution, Woods Hole, Massachusetts, 71 pp. plus appendixes.

IV

Oceanography Manpower Assessment Questionnaire (Academic Form)

1. Indicate the number of Ph.D. level staff (including post-docs) in your laboratory or department by sub-discipline from 1970-90.

	1970	1975	1980	1985	1990
B.O./M.B.	____	____	____	____	____
C.O./M.C.	____	____	____	____	____
MG&G	____	____	____	____	____
P.O.	____	____	____	____	____
O.E.	____	____	____	____	____

B.O./M.B.	=	Biological Oceanography/Marine Biology
C.O./M.C.	=	Chemical Oceanography/Marine Chemistry
MG&G	=	Marine Geology and Geophysics
P.O.	=	Physical Oceanography
O.E.	=	Ocean Engineering

2. Indicate the number of Ph.D. level staff by rank (Post-Doc, Assistant, Associate or Full Professors or equivalent research appointment) from 1970-90.

	1970	1975	1980	1985	1990
Post-Doc	____	____	____	____	____
Assist. Prof.	____	____	____	____	____
Assoc. Prof.	____	____	____	____	____
Full Prof.	____	____	____	____	____

3. Indicate the present age distribution of your Ph.D. level staff (i.e., faculty and staff with Principal Investigator status)

Number of Ph.D.s

	B.O./M.B.	C.O./M.C.	MG&G	P.O.	O.E.
<30	____	____	____	____	____
30-40	____	____	____	____	____
40-50	____	____	____	____	____
50-60	____	____	____	____	____
>60	____	____	____	____	____

4. Indicate the average number of months/years of institutional salary support ("hard money") provided to your Ph.D. level staff, including research series for each rank or its equivalent for 1990 for the most recent year available. Please exclude administrative support.

	Months/year
Post-doctoral Fellow	_____
Assistant Professor	_____

Associate Professor   _____

Full Professor   _____

5. Estimate (if possible) the number of Ph.D. level positions (replacement and new) that you anticipate filing in the next five years. What percentage of these will be institutionally supported (in average months per year)?

	B.O./M.B.	C.O./M.C.	MG&G	P.O.	O.E.
Replacement	_____	_____	____	____	____
New	_____	_____	____	____	____
% Supported	_____	_____	____	____	____

6. Institution name (will be kept confidential)

_____

V

Oceanography Manpower Assessment Questionnaire (Federal Laboratory Form)

1. Indicate the number of Ph.D. level staff (including post-docs) in your laboratory or department by sub-discipline from 1970-90.

	1970	1975	1980	1985	1990
B.O./M.B.	____	____	____	____	____
C.O./M.C.	____	____	____	____	____
MG&G	____	____	____	____	____
P.O.	____	____	____	____	____
O.E.	____	____	____	____	____

B.O./M.B.	= Biological Oceanography/Marine Biology
C.O./M.C.	= Chemical Oceanography/Marine Chemistry
MG&G	= Marine Geology and Geophysics
P.O.	= Physical Oceanography
O.E.	= Ocean Engineering

2. Indicate the present age distribution of your Ph.D. level staff (i.e., faculty and staff with Principal Investigator status).

Number of Ph.D.s

	B.O./M.B.	C.O./M.C.	MG&G	P.O.	O.E.
<30	_____	_____	_____	_____	_____
30-40	_____	_____	_____	_____	_____
40-50	_____	_____	_____	_____	_____
50-60	_____	_____	_____	_____	_____
>60	_____	_____	_____	_____	_____

3. Estimate (if possible) the number of Ph.D. level positions (replacement) and new that you anticipate filling in the next five years.

	B.O./M.B.	C.O./M.C.	MG&G	P.O.	O.E.
Replacement	_____	_____	_____	_____	_____
New	_____	_____	_____	_____	_____

4. Laboratory name (will be kept confidential)

_____

Principal source of funding _____

VI

Institutional Respondents
to Manpower Survey

Benedict Estuarine Research Laboratory
Columbia University, Lamont-Doherty Geological Observatory
Dauphin Island Sea Lab, Marine Environmental Sciences
Duke University Marine Laboratory
Florida State University, Department of Oceanography
Harbor Branch Oceanographic Institute, Inc.
Louisiana Universities Marine Consortium (LUMCON)
Monterey Bay Aquarium Research Institute
Moss Landing Marine Laboratories
Mote Marine Laboratory
New Jersey Marine Sciences Consortium
North Carolina State University, Department of Marine, Earth
    and Atmospheric Science
Nova University Oceanographic Center
Old Dominion University, Department of Oceanography
Oregon State University, College of Oceanography
Roger Williams College, School of Science and Mathematics
Skidaway Institute of Oceanography
State University of New York-Stony Brook, Marine Sciences
    Research Center
Texas A&M University, College of Geosciences, Department of
    Oceanography
University of Alaska, Institute of Marine Science

University of California-San Diego, Scripps Institution of
   Oceanography
University of California-Santa Barbara, Geology Department
University of California-Santa Barbara, Marine Science Institute
University of Delaware, College of Marine Studies
University of Georgia, UGA-Marine Institute
University of Hawaii, Hawaii Institute of Geophysics
University of Maine, Center for Marine Studies
University of Maryland, Center for Environmental and Estuarine
   Studies
University of Miami, Rosenstiel School for Marine and
   Atmospheric Sciences
University of Michigan, Center for Great Lakes and Aquatic
   Sciences
University of New Hampshire
University of Puerto Rico, Department of Marine Sciences
University of Rhode Island, Graduate School of Oceanography
University of Southern Mississippi, Center for Marine Science
University of South Florida, Department of Marine Science
University of Texas at Austin, Institute for Geophysics
University of the Virgin Islands
University of Washington, College of Ocean and Fishery Sciences
Virginia Institute of Marine Science, School of Marine Science
Woods Hole Oceanographic Institution

VII

Federal Respondents to Manpower Survey

Department of the Army
- U.S. Army Corps of Engineers

Department of Commerce, National Oceanic and Atmospheric Administration
- Environmental Research Laboratories (Office of Ocean and Atmospheric Research)
- Geodetic Research Laboratory (National Ocean Service)
- National Marine Fisheries Service laboratories
- Marine Mammal Laboratory (separate data as opposed to first three)

Department of Energy
- Argonne National Laboratory
- Brookhaven National Laboratory
- Oak Ridge National Laboratory
- Pacific Northwest Laboratory [Battelle Marine Sciences Laboratory]
- Savannah River Laboratory

Department of the Interior
- U.S. Geological Survey

Department of the Navy
- Naval Research Laboratory (Code 1005)
- Office of Naval Research (Codes 11 and 12)

Department of Transportation
- Coast Guard Academy
- U.S. Coast Guard Research and Development Center (Groton)
(•) International Ice Patrol (3 Ph.D. principal investigations since 1983), Oceanographic Unit (closed 1982)

National Aeronautics and Space Administration
- Goddard Oceans and Ice Branch
- Headquarters

Index